T0341032

LEAN SIX SIGMA FOR SMALL AND MEDIUM SIZED ENTERPRISES

A Practical Guide

LEAN SIX SIGMA FOR SMALL AND MEDIUM SIZED ENTERPRISES
A Practical Guide

Jiju Antony · S. Vinodh · E. V. Gijo

CRC Press
Taylor & Francis Group
Boca Raton London New York

CRC Press is an imprint of the
Taylor & Francis Group, an **informa** business

CRC Press
Taylor & Francis Group
6000 Broken Sound Parkway NW, Suite 300
Boca Raton, FL 33487-2742

© 2016 by Taylor & Francis Group, LLC
CRC Press is an imprint of Taylor & Francis Group, an Informa business

No claim to original U.S. Government works

Printed on acid-free paper
Version Date: 20151124

International Standard Book Number-13: 978-1-4822-6008-3 (Hardback)

Library of Congress Cataloging-in-Publication Data

Names: Antony, Jiju, author. | Vinodh, S., author. | Gijo, E. V., author.
Title: Lean Six Sigma for small and medium sized enterprises: a practical guide / authors,
  Jiju Antony, S. Vinodh, and E.V. Gijo.
Description: Boca Raton : Taylor & Francis, 2016. | Includes bibliographical references
  and index.
Identifiers: LCCN 2015045752 | ISBN 9781482260083 (hard cover)
Subjects: LCSH: Six sigma (Quality control standard) | Lean manufacturing.
Classification: LCC TS156.17.S59 A58 2016 | DDC 658.4/013--dc23
LC record available at http://lccn.loc.gov/2015045752

Visit the Taylor & Francis Web site at
http://www.taylorandfrancis.com

and the CRC Press Web site at
http://www.crcpress.com

Dedication

This book is dedicated to Frenie, Evelyn, Janane, Gaurav, Jayasree, Vaishnav, Vismaya and our parents.

Contents

Preface

A lot of companies out there still believe that Lean Six Sigma (LSS) is only for the big multinational corporations. This book is written to refute that myth. Our research and experience with a number of small and medium sized enterprises (SMEs) have clearly indicated that LSS can equally work in this sector. Obviously, there will be a number of challenges and barriers in the successful deployment of LSS within SMEs. This book covers some of the fundamental challenges and the common pitfalls which can be avoided in the introduction and successful deployment of LSS in the context of SMEs. Unlike larger corporations which invest heavily in LSS Black Belt training followed by the execution of a number of strategic projects, we recommend senior managers in SMEs to develop a number of LSS Green Belts and Yellow Belts at the outset of the initiative and then select the most talented candidates to become Black Belts if needed.

Don't look at your LSS team as a bunch of firefighters. They are fire preventers. We recommend SMEs to have an LSS infrastructure with a number of Green Belts, Yellow Belts, LSS project champions and even sponsors. The Green Belts are expected to identify the most critical problems across the business and try to develop a number of projects which can tackle these critical issues. In *Lean Six Sigma for Small and Medium Sized Enterprises*, the authors will systematically take you through the application of Six Sigma methodology for problem solving. A separate chapter is dedicated to the most appropriate tools and techniques which can be useful in each stage of the methodology. We want to highlight the fact that it will take a great deal of effort and commitment to learn and apply LSS in any SME context. We have tried our best to minimise the amount of math and statistics involved within our approach, however, it is virtually impossible to teach LSS without any statistics. We encourage companies to invest in Minitab, a powerful statistical software system for LSS programs that helps everyone to make decisions on the graphs.

This book is intended primarily for senior managers, middle managers and people on the shop floor who are preparing to become LSS deployment champions, Green Belts and Yellow Belts. A champion needs to focus on the LSS roadmap along with the senior management team so that he or she can communicate the progress of the journey on a regular basis. In an SME context, it is important that we involve all employees in process improvement projects using LSS tools; this will help them to hone their skills in problem solving and establish a sense of empowerment. LSS is not a quick fix or flavour of the month or management fad, but rather it is a proven business strategy which can deliver bottom-line results and a world-class practice for making your business processes efficient and effective.

The authors felt that SMEs cannot afford to invest a lot in training and this is often considered to be one of the major barriers for not launching a Six Sigma or LSS initiative in many SMEs. The authors of this book provide such SMEs with a roadmap for implementing and deploying LSS, followed by six excellent case studies showing how LSS tools have been integrated into LSS methodology. This would encourage a number of SMEs to embark on the LSS journey rather than relying too much on consultancy businesses, which often fail to develop a critical mass of people with required skills and expertise on the subject. We have written this book with the following salient features:

- Readiness factors for the introduction of LSS in SMEs
- A roadmap for deployment of LSS in SMEs
- Basic and advanced tools of LSS which are most appropriate in the context of SMEs
- Lean and Six Sigma Metrics
- Case studies of LSS from a number of SMEs
- Essentials of Lean and Six Sigma
- LSS project selection
- Six Sigma problem-solving methodology

We firmly believe the applications of LSS in SMEs will continue to grow over the years and this book is very timely. It can be a very useful guide for the implementation and deployment of LSS. We encourage senior managers in SMEs to use this book for training LSS Green Belts and Yellow Belts or for self-study to master the tools and methodology of Six Sigma.

This book consists of nine chapters:

Chapter 1 is an introduction to SMEs covering the characteristics of SMEs, contribution of SMEs to the world economy and some of the critical and fundamental differences between SMEs and larger firms.

Chapter 2 gives readers a general introduction to Continuous Improvement (CI) initiatives in SMEs. It highlights the critical factors required for the successful deployment of CI and the role of leadership and sustainability for CI.

Chapter 3 provides an excellent commentary on LSS covering a number of topics such as Lean methodology, some of the common myths of Six Sigma, strengths and weaknesses of LSS, background to LSS and some of the challenges in the implementation of LSS.

Chapter 4 gives a roadmap showing the key phases in the introduction and successful development of LSS initiatives within an SME. The readiness factors for the successful introduction of LSS are detailed in this chapter as well as LSS implementation infrastructure.

Chapter 5 is devoted to metrics of Lean, Six Sigma and Lean Six Sigma.

Chapter 6 explains the five phases of the Six Sigma methodology and the key activities which take place within each phase of the methodology.

Chapter 7 provides the most powerful tools and techniques of LSS which can be used in an SME environment. The authors provide readers with useful guidelines showing where, when, why and how these tools/techniques should be used with an illustrative example wherever necessary.

Chapter 8 is dedicated to project selection and prioritization showing how projects should be selected and prioritised. The authors also provide some tips for making the project selection process successful.

Chapter 9 is a great resource for LSS case studies. This chapter covers case studies on both Lean and Lean Six Sigma. Each case study is carefully chosen to illustrate the power of the LSS methodology and the associated tools within each phase of the methodology.

Finally, we would like to thank all the readers who are using this book for the LSS journey and we wish the very best of luck with your endeavours.

Acknowledgements

As authors of this book, we have benefited from the advice and help of a number of people in its preparation. The motivation for the development of this book emanates from the work of the first author Jiju Antony, based on his articles entitled 'Six Sigma in Small and Medium Sized Enterprises', which appeared in the *International Journal of Quality and Reliability Management* in 2005, and 'Gearing Six Sigma into UK Manufacturing SMEs: Results from a Pilot Study', which appeared in the *Journal of the Operational Research Society* in 2008.

When it comes to people, unfortunately no list can ever be complete and someone will be omitted. We hope those we do not mention here specifically will excuse us. We are intellectually indebted to the many academics, research scholars and practitioners of Lean and Six Sigma topics whose writing has blazed new trails and advanced the discipline of Lean Six Sigma. We are most grateful to the reviewers of the proposal and sample chapters for their invaluable suggestions that guided our preparation of this book.

It is our sincere hope that by reading this book, you will find something new which will challenge your personal thoughts in a new way. Your suggestions and constructive feedback regarding the coverage and contents of the book will be taken into consideration and we will do our best to overcome any shortcomings in the future editions of this book.

We take this opportunity to thank our publisher CRC Press (Taylor and Francis Group) for helping us to get this book in the market. For all of the many people with CRC Press who have helped us – a big thank you. Finally, we would like to thank all our family and research group members for their moral support during the preparation of the various chapters of our book.

Authors

Jiju Antony is recognised worldwide as a leader in Six Sigma methodology for achieving and sustaining process excellence. He founded the Centre for Research in Six Sigma and Process Excellence (CRISSPE) in 2004, establishing the first research centre in Europe in the field of Six Sigma. Professor Antony has successfully completed a BE in mechanical engineering from the University of Kerala (South India), an MSc in industrial engineering from the National University of Ireland, Ireland, and a PhD in quality engineering from the University of Portsmouth, UK. He is a fellow of the Royal Statistical Society (UK), fellow of the Institute for Operations Management (UK), fellow of the Chartered Quality Institute (CQI) and a fellow of the Institute of Six Sigma Professionals, UK. He is a Certified Master Black Belt and has been involved with over 100 Lean and Six Sigma–related projects across manufacturing, service and public sector organisations. He has a proven track record for conducting internationally leading research in the field of quality management and Lean Six Sigma. He has authored over 275 journal and conference papers and 6 textbooks. He has generated over £10 million from various research projects funded by European Commission as well as local government funding bodies in the United Kingdom. He has published over 90 papers on Six Sigma topics and is considered to be one of the highest in the world for the number of Six Sigma publications. Professor Antony has trained over 1000 people over the past 10 years on Lean and Six Sigma topics from over 150 companies in the United Kingdom and abroad, representing 21 countries. He is currently coaching and mentoring over 20 Lean Six Sigma projects across various UK public sector organizations including NHS, city councils, Police Scotland and the university sector. He is a past editor of the *International Journal of Six Sigma and Competitive Advantage* and has served as the editor of the *International Journal of Lean Six Sigma* since 2010 and associate editor of the *TQM and Business Excellence Journal* since September 2015. He is the founder of the first international conference on Six Sigma in the United Kingdom, back in 2004, and is also the founder of the first international conference on Lean Six Sigma for Higher Education. He has been a keynote speaker for various conferences around the world and has been a regular speaker for ASQ's Annual Lean Six Sigma Conference in Phoenix, Arizona, since 2009. He is on the editorial board of eight international journals including *Quality and Reliability Engineering International, International Journal of Quality and Reliability Management, TQM Journal, International Journal of Productivity and Performance Management, Measuring Business Excellence* and *Managing Service Quality* and a regular reviewer for *International Journal of Operations and Production Management, International Journal of Production Research, Journal of Operational Research Society, IIE Transactions, European Journal of Operational Research* and *Production, Planning and Control*. Professor Antony has worked on a number of consultancy projects with several blue chip companies such as Rolls-Royce, Bosch, Parker Pen, Siemens, Ford, Scottish Power, Tata, Thales, Nokia, Philips, GE, Nissan, Diageo and a number of small and medium sized enterprises.

S. Vinodh is an assistant professor in the Production Engineering Department of the National Institute of Technology, Tiruchirappalli, Tamil Nadu. Dr. Vinodh completed his bachelor's degree in mechanical engineering at Bharathiar University, Coimbatore, India; and master's degree in production engineering and PhD in mechanical engineering at Anna University, Chennai, India. He was awarded a National Doctoral Fellowship for pursuing doctoral research by the All India Council for Technical Education, New Delhi, India, during 2006–2008. He was awarded a Highly Commended Paper Award and Outstanding Paper Award by Emerald Publishers, United Kingdom, in 2009 and 2011 respectively. He was the recipient of the 2010 Innovative Student Project Award from the Indian National Academy of Engineering, New Delhi, India. He has published/presented over 100 papers at various international journals/conferences. He is serving on the editorial advisory board of the *International Journal of Lean Six Sigma* and *Journal of Engineering, Design and Technology*. His research interests include agile manufacturing, Lean manufacturing, sustainable manufacturing, Lean Six Sigma, additive manufacturing, product development and multi-criteria decision-making.

E.V. Gijo has worked as a faculty member in the Statistical Quality Control and Operations Research Division of the Indian Statistical Institute, Bangalore, India for the last 17 years. Dr. Gijo has successfully completed an MSc in statistics at M.G. University (Kerala), an M. Tech. in quality, reliability and operations research at the Indian Statistical Institute (Kolkata) and a PhD in statistics at M.G. University (Kerala). He is a certified Six Sigma Master Black Belt and a certified Lead Auditor for ISO 9001 & 14001 systems. He has a rich experience of conducting training and consultancy services in various industries including the automobile, chemical, wind energy, electrical, pharmaceutical, software, information technology/

information technology enabled services, business process outsourcing, health-care, insurance and construction industries. He has trained more than 800 Black Belts and more than 2000 Green Belts in India and abroad. He has also mentored around 1200 quality improvement projects including Six Sigma, Taguchi methods and other allied areas. He has authored around 45 journal and conference papers. Out of these publications, 15 articles are related to Six Sigma and were published in leading international journals. He is currently involved in conducting training and providing consultancy services to various industries in India. Dr. Gijo is on the editorial board of the *International Journal of Lean Six Sigma* and is a regular reviewer for *The TQM Journal, International Journal of Quality & Reliability Management, International Journal of Advanced Manufacturing Technology, International Journal of Computer Integrated Manufacturing, International Journal of Production Research, Quality and Reliability Engineering International, Journal of the Operational Research Society, Journal of Manufacturing Technology Management, International Journal of Six Sigma and Competitive Advantage* and so on.

chapter one

Introduction to small and medium sized enterprises (SMEs)

1.1 Introduction

Small and medium-scale enterprises (SMEs) contribute significantly to the global economy. SMEs form the backbone of developing countries and represent the rapidly growing sector. SMEs' key role is to promote entrepreneurial focus and innovation, thereby ensuring competitiveness. SMEs enable several larger companies and have a significant contribution to the world economy. This chapter reviews the definition of SMEs and their economic share, analysing the characteristics of small firms and their comparison with large businesses.

1.2 Definition of SMEs

While formulating plans and policies for any organisation, it is of the utmost importance to know the nature and type of business you are dealing with. Businesses across the globe are categorised based upon their magnitudes with respect to their size, capacity, capital invested and workforce employed. On a broad scale, large companies are grouped as one type while all other businesses are grouped as another. Though all enterprises falling under the second group enjoy the same social status as per legal norms followed by most countries across the globe, their economic levels are substantially different. Based on the creation of better development plans for such organisations, they are further categorised based on their magnitudes as SMEs. Broad classification of enterprises has been further extended to include micro enterprises along with SMEs to form micro, small and medium enterprises (MSMEs), due to continuous research efforts in recent years into the nature and functioning of enterprises.

Various regional authorities have classified SMEs based on vital criteria such as size of the plant, capacity, workforce size, investment and returns, which play a major role in deciding on the amenities and services to be provided and enjoyed by the firm. Studies on SMEs in the past found these criteria to be the appropriate definition of SMEs.

The definition of SMEs, based on enterprise categories of the European Union (Classen et al. 2014), is based on factors such as headcount and turn-over in euro (EUR). A firm with <10 headcount, <EUR 2 million in turnover is defined to be a micro enterprise, whereas a firm with 10–49 headcount, EUR 10–49 million in turnover is defined to be small enterprise, and a firm with 50–250 headcount, ≥EUR 50 million in turnover is defined to be a medium enterprise. These are active definitions for MSMEs followed by various European economies.

Alarape (2007) defined Nigerian SMEs as enterprises with labour size of 11–100 employees or a total cost of ≥50 million in Nigerian naira, which includes working capital and excludes land cost. Shehab (2008) defined Libyan SMEs as organisations having an employee count ranging from 50 to 250, making turnover in the range of 2–12 million in Libyan dinar (LYD) and LYD 1–8 million of turnover on the balance sheet. Du Toit et al. (2009) defined South African SMEs as enterprises which satisfy one or more of the following conditions: fewer than 200 employees, annual turnover of less than 64 million rand (ZAR), capital assets of less than ZAR 10 million and direct managerial involvement by owners.

There is more than just one definition for SMEs in Malaysia. The mean-ings based on several factors as described by Husin and Ibrahim (2014), which include SMEs based on sales turnover as published in the SME Performance Report and full-time employment as cited in the SME Annual Performance Report 2013/14, are that micro businesses are those having fewer than five employees in both the manufacturing sector, which con-tributes less than 250,000 Malaysian ringgit (MYR), and the services sector, which contributes less than MYR 200,000 to the nation's economy. Similarly, small businesses are those having employee counts within the range of 5–50 in the manufacturing sector, which contributes in the range from MYR 250,000 to less than MYR 10 million, and 5–19 employees in the service sec-tor, which contributes in the range from MYR 200,000 to less than MYR 1 million. Medium businesses are those having employee counts within the range of 51–150 in the manufacturing sector, which contributes in the range of MYR 10 million to MYR 25 million, and 20–50 employees in the services sector, which contributes about MYR 1 million to MYR 5 million.

Olusegun (2012) defines SMEs as firms with engagement in one form of business or another. The author states that definitions of SMEs differ across countries, industries, number of employees and asset value.

In the UK, the number of SMEs is estimated to be around 4.9 million. An enterprise in the UK context is termed as an SME if it has a turnover less than £25 million and the number of employees engaged is fewer than 250 (Maurya et al. 2015). About 99.3% of private sector businesses in the UK are SMEs with an annual turnover of £1.6 trillion that contribute 47% of the private sector turnover. In order to support SMEs, about 25% of the UK government's funding goes directly to SMEs.

Grover and Suominen (2014) define SMEs in the US as firms with fewer than 500 employees. SMEs are considered to be the backbone of the US economy. In the US, there are around 28 million SMEs functioning, and they are responsible for 34% of US export revenues. About 99% of the private firms that operate in the US are SMEs, and they employ over 50% of private sector employees (Parnell et al. 2015). They produce around 65% of the net new jobs in the private sector. Around 543,000 new business firms are getting started each month in the US. Only 25% of the SMEs around the US sustain for more than 15 years, and nearly 70% of the SMEs fail to survive for more than 2 years (Williams 2014).

In Canada, there are around 1.2 million SMEs, and they represent 54% of jobs generated by private sector businesses. SMEs in Canada represent 54.3% of the economic output generated by the business sector (Sui and Baum 2014). About 98.1% of Canadian SMEs have fewer than 100 employees, and about 55% of SMEs employ fewer than four employees.

In Australia, SMEs are defined as any business or company with revenues less than 20 million Australian dollars (Chong 2014). There are over 1.2 million SMEs in Australia, and they contribute 96% of all businesses and generate nearly 33% of the country's gross domestic product (GDP). About 63% of total employees engaged in the Australian private sector are from SMEs. The annual turnover of SMEs contributes to 9% of the Commonwealth revenue in Australia.

In South Africa, companies with fewer than 200 employees are termed as SMEs. They contribute to 91% of the formalised business and produce about 60% of the labour workforce. Their total economy contributes nearly 34% of GDP.

1.3   SMEs' contribution to world economy

SMEs account for 90% of businesses across the globe and 50%–60% in terms of employment (Jenkins 2004; Sannajust 2014). SMEs are major contributors to the social and economic development of Poland, since Poland has SMEs employing over 6 million people and generating 50% of its GDP. Of the total operating companies in Poland, 99.8% are SMEs (Walczak and Voss 2013). Of Dutch firms, 70.7% are small sized (10–49 employees), and 29.3% of firms are medium sized (50–250 employees) (Kraus et al. 2012).

In the beginning of 2014, 5.2 million small businesses accounted for 48% (12.1 million) of the UK private sector. The cumulative annual turnover of small businesses is £1.2 trillion, representing 33% of private sector turnover. In the beginning of 2014, 31,000 medium-sized businesses existed. These businesses contributed an annual turnover of £480 billion and employed 3.1 million people. In the context of the UK private sector, 5.2 million SMEs accounted for 60% (15.2 million). The cumulative annual turnover of SMEs is £1.6 trillion, representing 47% of turnover in the private sector (White 2014).

In the European Union context, SMEs are economically significant, with 99% of an estimated 23 million enterprises defined as SMEs. European SMEs generate 65 million jobs. Among those all are small enterprises, with 23 million companies (96.8%) which have fewer than 10 staff members and only 75,000 enterprises which employ more than 250 employees (Wach 2014). In the context of the European private sector, about two-thirds of jobs pertain to SMEs and generate 59% of added value.

In the world scenario, China has the highest number of SMEs in the world with 50 million. India is in second place with 48 million SMEs. SMEs in India contribute about 40% of its employment, 45% of its manufacturing output and 17% of the country's GDP (Malini 2013). SMEs are a vital contribution to the national economy of Malaysia. SMEs serve as a technology facilitator to industries and encompass 96.1% of total Malaysian firms (Hilmi and Ramayah 2009).

SMEs represent the backbone of the US economy. They form about 99% of all firms in the US, and 50% of the country's private sector employees are employed by SMEs. SMEs generate 65% of net new jobs in the private sector. They contribute over half of US non-farm GDP and form 98% of all US exports. SMEs contribute to 34% of US export revenue (Grover and Suominen 2014).

Over the last 12 years, the Mexican government has strengthened SMEs and its entrepreneurship policy framework and budget. This move has resulted in extensive improvements in the business environment for SMEs and entrepreneurship in Mexico. Mexico has an expanding SME sector. They account for 99.8% of enterprises and 72.3% of employment in the country. Micro enterprises (employing fewer than 10 people) of Mexico account for 96.1% of all businesses (OECD 2013). Brazil presently has around 6.3 million SMEs, and its annual operational revenue is US $39 million. SMEs in Brazil contribute 20% to the country's GDP and create around 47,000 new jobs, representing 52% of the formal jobs (Cravo et al. 2012).

Indonesian SMEs represent more than 90% of all firms external to the agricultural sector and also form the highest source of employment (Tambunan 2007). SMEs in Russia are crucial for its socio-economic prosperity, growth, employment and technical innovations. There are six SMEs per thousand people in Russia, 45 in the EU, 49.6 in Japan and 74.2 in the US (Zhuplev 2009).

India has nearly 36 million units of SMEs, which account for almost 50% of industrial output and 45% of the country's total exports (Nayak et al. 2014). Indian SMEs contribute about 8% to the country's present GDP and have significant opportunities to facilitate industrial growth across India. The Micro, Small and Medium Enterprises Development Act was framed by the government of India to address policy issues that affect the financial performance of SMEs and thereby enhance their competitiveness. As per the new Micro, Small and Medium Enterprises Development Act 2006, the number of Indian SMEs ranges from 7.8 million to 13 million. The total share in GDP of SMEs in India is more than 80%, and more than 90% of all the enterprises are SMEs.

1.4   Characteristics of SMEs

SMEs are being classified based on their size and structure, which includes employee headcount, assets and financial turnover that is below a particular limit. SMEs are the key drivers of growth and are important for efficient and competitive markets to facilitate job opportunities (Beck and Demirguc-Kunt 2006). The difference between an SME and a large enterprise originates from how the business entity is conceptualised and managed. SMEs might have a flat hierarchy and an integrated set of business functions, whereas large enterprises have a matrix organisation and a set of business units which focus on specific market segments and business domains. Business ownership patterns are different between the two entities (Hoffmann and Schlosser 2001). SMEs must evolve the focus of their enterprises in a manner compatible with the phases of the industries they serve. The roles of technology evolve in a similar manner (Hallberg 2000).

1.4.1   Low start-up costs

Start-up costs for small-scale businesses are normally lower, pertaining to specific business models and the kind of products or services provided (Blair and Marcum 2015). This contrasts with larger businesses requiring huge investments during the start-up stage.

1.4.2   Portability

A small-scale business is normally portable, enabling set-up and tear-down (Simatele 2014). Small-scale businesses also require an avenue to accept payments such as small credit card terminals.

1.4.3   Leadership

An SME's success often relies on its owner's leadership skills, characterised by unity of ownership, management, liability and risk (Aslan et al. 2011). Due to flat organisational structure and limited resources, the ownership and management of day-to-day operational activities rests with the owner/leader in SMEs compared to larger organisations where leadership is shared, dispersed and institutionalised.

1.4.4   Management structure

The management in small businesses is mostly through direct supervision or supervision by owner(s)/manager(s) as compared to delegation or decentralisation of responsibilities to employees as witnessed in large firms. Understanding the management of small firms is an essential component in perceiving the relationship among ownership and decision-making, managerial styles, organisational structure and culture and business development pattern (Walczak 2005). SMEs have flat organisational structure, i.e. fewer levels of management, and fewer departmental interfaces promoting a flexible work environment which facilitates rapid communication, fast decision-making process and speedier implementation.

1.4.5   Planning

SMEs exist in uncertain and dynamic environments, where innovation, flexibility and responsiveness may be vital for survival (Wang et al. 2007). Research studies have emphasised that SMEs cannot survive or sustain without strategic planning and strategic developments (Dibrell et al. 2014). In such firms, the strategic process is streamlined as in many large firms.

1.4.6   Systems and procedures

The extent of formal systems in small firms ranges from minimal/non-existent to basic levels compared to large firms, where they have mature and formally established systems and procedures in place (Terziovski 2010). The simple processes in SMEs allow flexibility and encourage innovation and a speedy response to customer requirements.

1.4.7   Human resources

Usually, very few employees are engaged for a small-scale business. During busy periods, business may have one or two employees (Thomas and Webb 2003). Small-scale businesses are staffed by owners themselves to retain profits. It is easier in the SME environment to educate and train employees due to the smaller workforce compared to their larger counterparts. But SMEs face financial constraints, and they cannot mobilise the same resources for education and training that larger firms can (Nolan and Garavan 2015). Training and staff development in SMEs is mostly ad hoc and small-scale compared to the scheduled and large-scale training in larger firms.

1.4.8   Market and customer focus

Small-scale businesses require a much smaller area than larger corporations or private businesses, and they focus on single communities (Keskin

2006). SMEs are very close to their products and customers, allowing rapid information flow between customers and the company and creating a higher sense of responsibility (Sain and Wilde 2014). SMEs usually depend on few customer bases with limited external contacts. On the contrary, larger firms have a larger span of activities dispersed externally with a large customer domain.

1.4.9   Operational improvement

Implementation of best-in-class management practices such as total quality management (TQM), Lean, Six Sigma and Kaizen in large organisations is well documented and cited in literature (Gunasekaran et al. 2000). However, existing evidence suggests that SMEs are slower to adopt such formalised management practices due to lack of understanding of these management practices, lack of resources (people, finance, time, etc.) and lack of knowledge and short-term strategic planning.

1.4.10   Innovation

Innovation is a key source of competitive advantage for organisations (Lee et al. 2010). The relative advantage of large firms lies mostly in utilising their resources and good external networking, while SMEs are relatively strong in innovation where effects of scale are not important as compared to their behavioural attributes. In SMEs, incremental innovation refers to improving or revising the existing features of a product or service, and radical innovation involves developing products or services that offer wider benefits to customers than the existing products and services (Klewitz and Hansen 2014). Most of the SMEs focus on incremental innovation because the necessity for product enhancement is less and due to the unavailability of technological resources (Prajogo and McDermott 2014). When undergoing radical innovations, the firm must improve its learning and knowledge-sharing capability and must develop appropriate infrastructure. Moreover the decision to go with incremental or radical innovation is dependent on the SME's willingness and its absorptive capabilities (Massa and Testa 2008).

1.4.11   Networking

Networking and alliances play a vital role in the skill development of entrepreneurs for small firms and provide them a chance to build strategic market positions that contribute to competitive advantage (Gilmore et al. 2006). Research has shown that SMEs are better able to innovate when they are part of clusters. This is the reason most advanced manufacturing concepts are implemented in a cluster approach wherein groups of SMEs are brought together with similar product manufacturing, and it promotes inclusive growth.

1.4.12   Revenue and profitability

Revenue for small-scale business is generally less than for companies working on a wider scale (Esselaar et al. 2006). Established small-scale businesses often establish their facilities and equipment to maintain lower costs than more leveraged businesses.

1.4.13   Ownership and taxes

The corporate form of business organisations is not adaptable to small-scale operations (Smallbone and Welter 2001). Small-scale businesses prefer to organise as sole proprietorships, partnerships or limited liability companies.

1.4.14   Locations

A small-scale business is performed in a limited area. These companies do not have sales outlets in multiple states or countries due to financial constraints and limited orders. The location of SMEs plays a major role in deciding a firm's direct and indirect expenses. SMEs located in metropolitan areas have lower cost of access to foreign markets and can promote exports due to the connectivity which SMEs in remote locations lack (Freeman and Styles 2014). SMEs located in remote places are mostly affected by circumstances inherent to their location that prevent them from enhancing business performance.

1.5   SMEs versus larger firms

Global markets are dominated by larger firms due to their huge economies of scale and promotional expenses, which SMEs lack. Even though SMEs are not able to compete with larger firms in terms of capital investments and economies of scale, they have a strong potential to compete on service and value metrics with larger firms. SMEs are more adaptable and flexible in satisfying customer needs, which most larger firms fail to do. There are key factors which need to be identified when benchmarking SMEs and larger firms. Based on a literature review, the following factors are found to be vital to compare SMEs with larger firms. Innovation, speed of attitude towards risk, decision-making, resource allocation, understanding and management of business models are some of the key factors used to differentiate small and medium enterprises and large businesses.

1.5.1   Innovation

The business scenario experiences a turbulent situation due to rapidly changing customer preferences and the emergence of newer technologies. Organisations need to make appropriate investments on products or services which will withstand competitiveness that is not possible with SMEs. Due to increasing competition, both larger firms and SMEs are expected to be innovative. The extent of innovation culture varies from larger firms to SMEs. Large firms experience a high degree of innovation whereas SMEs are expected to be reasonably innovative.

1.5.2   Attitude towards risk

Large-scale enterprises undertake higher risk in executing business tasks. This is because their capital is larger, and buffers exist to absorb any uncertainties. However, SMEs must consider the negative consequences that may result due to investment and revenue generation.

1.5.3   Decision-making

Large enterprises with higher levels in the organisational hierarchy will often require more time to make decisions, which creates frustration when immediate decisions are needed. Any delay in the decision-making process may affect the progress of the company. In the context of SMEs with a lower hierarchical structure, decisions can be made faster in vital situations.

1.5.4   Resource allocation

In small businesses, resources are scarce, and the allocation is based on bottom-line benefits as they do not have streamlined resource allocation procedures. This scenario may not be so distinct in a larger corporation wherein appropriate procedures will be followed for the planning and allocation of resources.

1.5.5   Understanding and management of business models

A large enterprise establishes business models in a comprehensive manner when compared with SMEs. Larger firms conduct periodic training and focus on activities to attain the company's goal, which is not the case in SMEs. Most SMEs do not have a comprehensive business model with interlinked functions.

1.6   Summary

The world market witnesses globalisation as a result of which larger firms outsource their business to smaller firms. As a result, SMEs are being recognised worldwide and significantly contribute to economic development, job opportunities and economic welfare. Despite contributing towards the worldwide economy, SMEs also face some challenges in the global market that restrict them from being competitive. Some of the challenges faced by SMEs in global markets are low levels of expenditure on research and development (R&D) activities, lack of modern information and technological systems, poorly developed innovation systems, stringent policies, lack of knowledge on marketing and branding, low managerial capabilities and lack of contacts with major local and international enterprises (Khalique et al. 2011; Rowley and Cooke 2014). These challenges can be overcome by providing a knowledge-based understanding to the SMEs on improving R&D infrastructure and to enhance the management's focus on intellectual capital investments that should be spent on physical assets. SMEs must learn to improve their marketing channels and concentrate more on product innovation for sustaining themselves in the global market. Based on the contribution of small enterprises, their importance needs to be completely understood. This chapter reviewed the definitions of SMEs, the contribution of SMEs towards the global economy and the characteristics of SMEs compared to larger firms. The vital differences between large and small organisations needs to be analysed based on the characteristics of small firms and their differentiators with large organisations. This kind of distinction is needed as the business models and theories developed for large organisations cannot be directly applied for smaller organisations. The difference is based on how the small firms run their businesses and the frameworks for facilitating continuous improvement.

References

Alarape, A. A. (2007). Entrepreneurship programs, operational efficiency and growth of small businesses. *Journal of Enterprising Communities: People and Places in the Global Economy* 1(3): 222–239.

Aslan, Ş., Diken, A. and Şendoğdu, A. A. (2011). Investigation of the effects of strategic leadership on strategic change and innovativeness of SMEs in a perceived environmental uncertainty. *Procedia: Social and Behavioral Sciences* 24: 627–642.

Beck, T. and Demirguc-Kunt, A. (2006). Small and medium-size enterprises: Access to finance as a growth constraint. *Journal of Banking & Finance* 30(11): 2931–2943.

Blair, E. S. and Marcum, T. M. (2015). Heed our advice: Exploring how professionals guide small business owners in start-up entity choice. *Journal of Small Business Management* 53(1): 249–265.

Chong, S. (2014). Business process management for SMEs: An exploratory study of implementation factors for the Australian wine industry. *Journal of Information Systems and Small Business* 1(1/2): 41–58.

Classen, N., Carree, M., Van Gils, A. and Peters, B. (2014). Innovation in family and non-family SMEs: An exploratory analysis. *Small Business Economics* 42(3): 595–609.

Cravo, T. A., Gourlay, A. and Becker, B. (2012). SMEs and regional economic growth in Brazil. *Small Business Economics* 38(2): 217–230.

Dibrell, C., Craig, J. B. and Neubaum, D. O. (2014). Linking the formal strategic planning process, planning flexibility, and innovativeness to firm performance. *Journal of Business Research* 67(9): 2000–2007.

Du Toit, G. S., Erasmus, B. J. and Strydom, J. W. (2009). *Definition of a Small Business and Introduction to Business Management.* 7th edition. Cape Town: Oxford University Press.

Esselaar, S., Stork, C., Ndiwalana, A. and Deen-Swarray, M. (2006, May). ICT usage and its impact on profitability of SMEs in 13 African countries. In *Information and Communication Technologies and Development, 2006. ICTD '06. International Conference on IEEE.* (pp. 40–47).

Freeman, J. and Styles, C. (2014). Does location matter to export performance? *International Marketing Review* 31(2): 181–208.

Gilmore, A., Carson, D. and Rocks, S. (2006). Networking in SMEs: Evaluating its contribution to marketing activity. *International Business Review* 15(3): 278–293.

Grover, A. and Suominen, K. (2014). 2014 Summary: State of SME finance in the United States. White Paper.

Gunasekaran, A., Forker, L. and Kobu, B. (2000). Improving operations performance in a small company: A case study. *International Journal of Operations & Production Management* 20(3): 316–336.

Hallberg, K. (2000). A market-oriented strategy for small and medium scale enterprises. Vol. 63. White Paper. Washington, DC: World Bank Publications.

Hilmi, M. F. and Ramayah, T. (2009). Market innovativeness of Malaysian SMEs: Preliminary results from a first wave data collection. *Asian Social Science* 4(12): 42.

Hoffmann, W. H. and Schlosser, R. (2001). Success factors of strategic alliances in small and medium-sized enterprises: An empirical survey. *Long Range Planning* 34(3): 357–381.

Husin, M. A. and Ibrahim, M. D. (2014). The role of accounting services and impact on Small Medium Enterprises (SMEs) performance in manufacturing sector from east coast region of Malaysia: A conceptual paper. *Procedia-Social and Behavioral Sciences* 115: 54–67.

Jenkins, H. (2004). A critique of conventional CSR theory: An SME perspective. *Journal of General Management* 29(4): 55–75.

Keskin, H. (2006). Market orientation, learning orientation, and innovation capabilities in SMEs: An extended model. *European Journal of Innovation Management* 9(4): 396–417.

Khalique, M., Isa, A. H. B. M., Shaari, N., Abdul, J. and Ageel, A. (2011). Challenges faced by the small and medium enterprises (SMEs) in Malaysia: An intellectual capital perspective. *International Journal of Current Research* 3(6): 398–401.

Klewitz, J. and Hansen, E. G. (2014). Sustainability-oriented innovation of SMEs: A systematic review. *Journal of Cleaner Production* 65: 57–75.

Kraus, S., Rigtering, J. C., Hughes, M. and Hosman, V. (2012). Entrepreneurial orientation and the business performance of SMEs: A quantitative study from the Netherlands. *Review of Managerial Science* 6(2): 161–182.

Lee, S., Park, G., Yoon, B. and Park, J. (2010). Open innovation in SMEs: An intermediated network model. *Research Policy* 39(2): 290–300.

Malini, G. (2013). SMEs employ close to 40% of India's workforce, but contribute only 17% to GDP, http://articles.economictimes.indiatimes.com/2013-06-09/news/39834857_1_smes-workforce-small-and-medium-enterprises (Retrieved 10 September 2015).

Massa, S. and Testa, S. (2008). Innovation and SMEs: Misaligned perspectives and goals among entrepreneurs, academics, and policy makers. *Technovation* 28(7): 393–407.

Maurya, U. K., Mishra, P., Anand, S. and Kumar, N. (2015). Corporate identity, customer orientation and performance of SMEs: Exploring the linkages. *IIMB Management Review* 27(3): 159–174.

Nayak, R., Kumar, A. and Sengupta, R. (2014). Barriers affecting implementation of Technology Transfer (TT) in apparel manufacturing Indian SMEs. *International Journal of Applied Sciences and Engineering Research* 4(4): 417–426.

Nolan, C. T. and Garavan, T. N. (2015). Human resource development in SMEs: A systematic review of the literature. *International Journal of Management Reviews*.

OECD (2013). *Mexico: Key Issues and Policies, OECD Studies on SMEs and Entrepreneurship.* OECD Publishing, Paris, http://dx.doi.org/10.1787/9789264187030-en.

Olusegun, A. I. (2012). Is small and medium enterprises (SMEs) an entrepreneurship? *International Journal of Academic Research in Business and Social Sciences* 2(1): 487.

Parnell, J. A., Long, Z. and Lester, D. (2015). Competitive strategy, capabilities and uncertainty in small and medium sized enterprises (SMEs) in China and the United States. *Management Decision* 53(2): 402–431.

Prajogo, D. and McDermott, C. M. (2014). Antecedents of service innovation in SMEs: Comparing the effects of external and internal factors. *Journal of Small Business Management* 52(3): 521–540.

Rowley, C. and Cooke, F. L. (2014). *The Changing Face of Management in China.* New York: Routledge.

Sain, S. and Wilde, S. (2014). *Soft Skills within Customer Knowledge Management and their Impact on Customer Focus.* Berlin: Springer International.

Sannajust, A. (2014). Impact of the world financial crisis to SMEs: The determinants of bank loan rejection in Europe and USA. Working Paper 327.

Shehab, M. A. (2008). Factors influencing accounting information system performance among small and medium enterprises (SMEs) in Tropli. MSc thesis. Universiti Utara Malaysia.

Simatele, M. (2014). Enhancing the portability of employability skills using e-portfolios. *Journal of Further and Higher Education* 39(6): 862–874.

Smallbone, D. and Welter, F. (2001). The role of government in SME development in transition economies. *International Small Business Journal* 19(4): 63–77.

Sui, S. and Baum, M. (2014). Internationalization strategy, firm resources and the survival of SMEs in the export market. *Journal of International Business Studies* 45(7): 821–841.

Tambunan, T. (2007). Entrepreneurship development: SMES in Indonesia. *Journal of Developmental Entrepreneurship* 12(1): 95–118.

Terziovski, M. (2010). Innovation practice and its performance implications in small and medium enterprises (SMEs) in the manufacturing sector: A resource-based view. *Strategic Management Journal* 31(8): 892–902.

Thomas, A. J. and Webb, D. (2003). Quality systems implementation in Welsh small-to medium-sized enterprises: A global comparison and a model for change. *Proceedings of the Institution of Mechanical Engineers, Part B: Journal of Engineering Manufacture* 217(4): 573–579.

Wach, K. (2014). The scale of internationalisation and Europeanisation of SMEs and their functioning in the spatial systems of the European Union. *Przedsiębiorczość-Edukacja* 10: 136–147.

Walczak, D. and Voss, G. (2013). New possibilities of supporting Polish SMEs within the Jeremie initiative managed by BGK. *Mediterranean Journal of Social Sciences* 4(9): 759.

Walczak, S. (2005). Organizational knowledge management structure. *The Learning Organization* 12(4): 330–339.

Wang, C., Walker, E. A. and Redmond, J. L. (2007). Explaining the lack of strategic planning in SMEs: The importance of owner motivation. *International Journal of Organisational Behaviour* 12(1): 1–16.

White, S. (2014). Business population estimates for the UK and regions (2011), https://www.gov.uk/government/uploads/system/uploads/attachment_data/file/377934/bpe_2014_statistical_release.pdf (Retrieved 10–9–2015).

Williams, D. A. (2014). Resources and failure of SMEs: Another look. *Journal of Developmental Entrepreneurship* 19(1): 1450007.

Zhuplev, A. (2009). Small business in Russia: Trends and outlook. Baltic Rim. White Paper.

chapter two

Continuous improvement initiatives in SMEs

2.1 What is continuous improvement?

The concept of continuous improvement (CI) comes from the Japanese term *Kaizen* that was initially developed and propagated by Masaaki Imai, who is recognised as the father of CI. Kaizen is a compound word in Japanese that includes two concepts: kai (change) and zen (to improve). Boer and Gertsen (2003) provide us with a very meaningful definition of CI and define CI as 'the planned, organised and systematic process of ongoing, incremental and company-wide change of existing practices aimed at improving company performance'.

CI is an ongoing effort to improve products, services or processes. These efforts can seek 'incremental' improvement over time or 'break-through' improvement all at once. It is based upon a belief that continual improvement can be brought about by a never-ending series of small changes. It is important to note that all employees must participate in a CI initiative, and senior management must be totally engaged with the initiative right from the outset.

CI doesn't happen overnight – it is a never-ending journey that requires long-term vision and uncompromising commitment from the senior management team across the organisation early in the process. The senior management team should convey the message that CI is not just a cost-cutting exercise; rather it is about changing the culture of the organisation. In order to weave CI into the DNA, organisations should focus on some key principles of CI, such as leadership, employee engagement and process improvement metrics for CI, a data-driven approach to improvement and robust governance.

How do we make sure that a corporate culture is ready for a CI initiative? First of all, leaders in an organisation should understand that CI is a philosophy and not a set of tools and techniques for problem solving. Many senior managers are often demanding a quick-fix solution to their problems, which can deliver short-term results. Moreover, to successfully build a culture for CI and other change initiatives, people in the organisation need to be engaged and be a part of the change process. Leaders should work with the CI team to set clear goals, but they should give their team the freedom to make their own decisions to achieve the objectives. It is important that each team member feels valued and respected, even when his or her point of view or approach is not adopted. In addition to the above, it is strongly recommended for leaders to link the work of the team to business goals, by being decisive and clear in prioritising which CI projects will deliver the greatest value.

2.2　Continuous improvement practices in small and medium sized enterprises (SMEs)

The available literature debates the use of CI practices in the context of SMEs. One of the reasons why SMEs should work on CI is that they often serve as suppliers to large companies which demand CI and demonstrate the various capabilities to meet their requirements. In this section, we will briefly talk about different CI initiatives adopted by SMEs. These CI initiatives include Total Quality Management (TQM), Lean and Six Sigma.

Ghobadian and Gallear (1996) reported on the case studies of four SMEs where they investigated the reasons for adopting TQM, the main steps involved in implementation, the impact and changes resulting from its adoption and the difficulties in implementation. They concluded that SMEs could apply TQM with considerable success, and they pointed out the strengths inherent in SMEs which were beneficial for this.

Shea and Gobeli (1995) in their article cite some of the motives in the implementation of TQM reported by a group of small companies they studied. They are

1. Promotion of growth – it is easier to convince the company's bankers to invest in it if there is evidence that the organisation is well run.
2. Management's belief in the principle of customer satisfaction and employee empowerment, which reflects the management style supporting TQM.
3. Changing customer expectations, even for organisations seen to be doing well (competitive issue).
4. Making work more enjoyable.
5. Improving poor company performance if the company is not doing well (survival issue).

So what are the key lessons learned from the implementation of TQM in SMEs? Certain fundamental processes necessary for TQM implementation exist, but the rate at which an organisation carries them out depends very much on the level of resources available. This means that SMEs should adopt TQM in a much more staggered and progressive manner. SMEs cannot afford massive investments such as a whole year on TQM education alone, or on visits to TQM companies to perform benchmarking. Small businesses should, on a small scale, consider improvement projects that can reduce costs, increase profits, reduce rejects and reduce failures, with a high degree of success, in the shortest possible time. TQM will not solve every problem, and in fact, it may create others. It is just one of the many philosophies that a company can adopt to achieve its business goals; it is not the panacea for all ills.

According to Achanga et al. (2006), although Lean manufacturing is becoming a popular technique for productivity improvement, SMEs are still not certain of the cost of its implementation and the likely tangible and intangible benefits they may achieve. Most of these companies fear that implementing Lean manufacturing is costly and time consuming. Moreover, effective application and utilisation of Lean manufacturing within SMEs will be delayed or may not be achieved at all unless SMEs restructure their focus to become more receptive and capable of absorbing new ideas.

A study carried out by Dora et al. (2013) shows that some Lean manufacturing practices such as total productive maintenance (TPM), employee involvement and customer involvement are widely used among the European food SMEs in comparison with other Lean practices. The finding on the use of TPM by SMEs is not in line with the literature, which states that the uptake of TPM is very slow in SMEs. The result of this study also illustrates that food-processing SMEs benefit from implementing Lean manufacturing practices, particularly regarding reducing cost, improving profitability, increasing productivity and reducing customer complains. The full benefits of Lean manufacturing are not realised by the food SMEs because of their early stage of adoption.

Wessel and Burcher (2004) in their study identify the specific requirements for the implementation of Six Sigma based on a sample of SMEs in Germany. This study also examines how Six Sigma has to be modified to be applicable and valuable in an SME environment. This is the first study of its kind to be carried out on Six Sigma in SMEs.

Snee and Hoerl (2003) argue that there is nothing inherent in Six Sigma that makes it more suitable for large companies. They also suggest that the greatest barrier to implementation in small companies to date has been the way the major Six Sigma training providers have structured their offerings. More recently, as more and more sets of deployment guides and training materials have become available, the pricing structures have begun to change. Today, it is much easier for SMEs to obtain good external resources without a large upfront payment. Adams et al. (2003) suggest that the initial focus of SMEs can be to reduce quality costs or waste in the system. Effort and investment, as well as results in smaller companies, are more visible within a short time.

Davis (2003) points out that the problem arises in SMEs when they solicit deployment proposals from Six Sigma consulting companies only to learn that the traditional Six Sigma implementation approach in large companies can require millions of dollars investment, dedication of their best people on Six Sigma projects and training of the masses. He also argues that using a yellow belt approach allows SMEs to implement Six Sigma at a less costly, more manageable pace. He adds that the 'one size fits all' Six Sigma black belt deployment model is not practical for every company.

A cross-sectional study was conducted by Antony (2015) to assess the current status of Six Sigma implementation in UK SMEs. The results of the study showed that many SMEs were not aware of Six Sigma or did not have the resources to implement Six Sigma projects. Management involvement and participation, linking Six Sigma to customers and linking Six Sigma to business strategy were the most critical factors cited for the successful deployment of Six Sigma in SMEs. This paper surveyed the use of Six Sigma in SMEs and showed that those who adopt it have reaped benefits at both strategic and operational levels. If these benefits are to increase, there needs to be greater dissemination of its benefits and the creation of user groups that support SMEs in sharing and exchanging experiences of successful deployment of Six Sigma, thus promoting the best-in-class practice within the user group.

2.3   Critical success factors in the implementation of CI practices in SMEs

Critical success factors (CSFs) are those factors which are critical to the success of any organisation, in the sense that, if objectives associated with the factors are not achieved, the organisation will fail – perhaps catastrophically so. Oakland (2000) defines CSFs as 'a term used to mean the most important sub goals of a business or organization … what must be accomplished for the mission to be achieved'. In the context of Six Sigma project implementation, CSFs represent the essential ingredients without which a project stands little chance of success.

Lee (2004) carried out an exploratory study in small Chinese manufacturing firms to investigate the present status of TQM and its perception and development in these small firms. The CSFs for TQM implementation in his study included top management commitment, employee participation, supplier involvement and training and education.

Assarlind and Gremyr (2014) identify a number of CSFs for CI initiatives based on a review of 59 papers. A number of factors are extracted and then grouped into six categories: contextualisation, gradual implementation using realistic goals, involvement and training of employees, involvement of external support, management involvement and fact-based follow-up. These factors are not all unique to SMEs, but collectively, they are uniquely targeted towards SMEs.

The following CSFs should be taken into account for the successful deployment of Six Sigma in SMEs: leadership and management commitment; organisational infrastructure; cultural change; education and training; vision and plan statement; linking Six Sigma to customer; linking Six Sigma to business strategy; linking Six Sigma to employees; linking Six Sigma to suppliers; communication; understanding of Six Sigma; project management skills; and project prioritisation and selection.

Antony (2015) has identified the CSFs for the successful implementation of Six Sigma within UK manufacturing SMEs. These include management involvement and participation, linking Six Sigma to customers, linking Six Sigma to the business strategy, organisational infrastructure, understanding of Six Sigma methodology, training on Six Sigma and project prioritisation and selection. Achanga et al. (2006) identify four CSFs for the successful adoption of Lean manufacturing within the SME environment. These include leadership and management, financial capabilities, organisational culture and skills and expertise.

2.4   Leadership for CI

Although the importance of CI as a priority in any business setting has been recognised by everyone regardless of the size and nature of the organisation, the leadership for achieving and sustaining CI has been a constant problem over the years. Juran et al. (1995) in one of their articles state that 'attaining quality leadership requires that senior managers personally take charge of the continuous improvement initiative'. The best example for this is explicitly demonstrated by the former CEO of Motorola, Robert Galvin, who has made a habit of making quality the very first item on the agenda of executive staff meetings.

So, what sort of leadership is required for an organisation to achieve and sustain a CI initiative? In our personal opinion, the following traits, characteristics or roles should be required in a leader for an organisation to sustain CI:

- Setting strategic and visionary leadership.
- Defining and communicating the strategy adopted by the organisation to achieve and maintain quality.
- Empowering employees and make them accountable for maintaining CI of their own work processes.
- Creating the power of an environment of trust, openness and honest communication.
- Creating an environment that promotes creativity, innovation and continual improvement.

- Inspiring, motivating and recognizing employee contributions.
- Developing challenging goals and targets – through goal setting, leaders are able to foster constant growth and development across the organisation, by continually setting realistic and measurable goals within each department.
- Creating CI projects across the organisation and ensuring that all employees actively participate in such projects. We recommend some good coaching and mentoring of projects with the help of team leaders and CI specialists within each business function in an organisation.

Perhaps the leadership style that most relates to quality leadership is transformational leadership, which searches for ways to help motivate followers by satisfying high order needs and more fully engaging them in the process of the work. Moreover, transformational leaders encourage quality improvement by creating a culture of trust, creating an inspirational vision focusing on quality, developing a culture that supports a paradigm shift in quality, etc. Deming (1986) stated that 'the required transformation of the Western style of management requires that managers be leaders'. Although we have described the roles or characteristics of leaders for sustaining CI in an organisation, we would argue that we still have not done much research on the skills that are required for creating the Lean Six Sigma (LSS) leaders of tomorrow in an SME environment.

2.5   Sustainability of CI initiatives

Over the last few decades, a number of scholars have studied how corporations can achieve a competitive advantage through LSS or any CI initiatives. In other words, LSS or similar business process improvement methodologies have been some of the primary weapons used by such corporations to win business in the global market place. Sustaining a competitive advantage in CI requires sustaining a high level of quality at low operational costs relative to competition. Sustaining a high level of quality entails meeting and exceeding customers' expectations over time. As customers' expectations change over time, it is important for organisations to adapt their critical processes too.

Research has shown that a number of factors are critical for sustaining LSS or CI over a period of time. These include commitment to and leadership on CI from the senior management team, planning and organisation for CI, continuous education and training on CI across the organisation at all levels, measurement and feedback, use of tools and techniques of CI and culture change. Some of the barriers to sustaining CI are fear and resistance to change, inadequate leadership, lack of resources for sustaining CI (for instance, no CI champions or a formal CI team) and so on. *A recent study shows that companies that practice effective change management are*

71% more likely to achieve and sustain their objectives than those that minimise the people side of change. Top that off with the fact that those same companies are 55% more likely to be on or ahead of schedule in their efforts to accomplish their business goals.

In the authors' view, organisations which advocate a dynamic capability approach will likely stay ahead in the race for quality in the forthcoming years. Dynamic capability is 'the firm's ability to integrate, build and reconfigure internal and external competences to address rapidly changing environments'. In a nutshell, we would highlight the fact that organisations which build a dynamic capability would sustain LSS over a period of time. Having pursued a number of research projects and delivered a number of consultancy assignments with many organisations, the authors have witnessed that only a handful of companies are sustaining CI initiatives over time, and many companies do not have sustainability frameworks when it comes to CI methodologies such as Lean, Six Sigma or Lean Six Sigma.

2.6   Summary

This chapter begins with a basic introduction to CI and its importance in organisations. It then presents the various CI practices adopted by SMEs, which include TQM, Lean and Six Sigma. The chapter then goes on to talk about the CSFs for the implementation of any CI initiatives within the SME context. The authors then clearly highlight some of the key traits that leaders in SMEs should possess if they are to be successful with the CI initiative. The last part of the chapter reveals some of the challenges in sustaining the CI initiative and the significance of dynamic capability to sustain any CI initiative over a period of time.

References

Achanga, P., Shehab, E., Roy, R. and Nelder, G. (2006). Critical success factors for lean implementation within SMEs. *Journal of Manufacturing Technology Management* 17(4): 460–471.

Adams, C. W., Gupta, P. and Wilson, C. (2003). *Six Sigma Deployment*. Burlington, MA: Butterworth-Heinemann.

Antony, J. (2015). The ten commandments of quality: A performance perspective. *International Journal of Productivity and Performance Management* 64(5): 723–735.

Assarlind, M. and Gremyr, I. (2014). Critical factors for quality management initiatives in small and medium sized enterprises. *Total Quality Management & Business Excellence* 25(3/4): 397–411.

Boer, H. and Gertsen, F. (2003). From continuous improvement to continuous innovation, a retroperspective. *International Journal of Technology Management* 26(8): 805–827.

Davis, A. (2003). Six Sigma for small companies. *Troy* 42(11): 20.

Deming, W. E. (1986). *Out of the Crisis.* Cambridge, MA: MIT Press.

Dora, M., Kumar, M., Van Goubergen, D., Molnar, A. and Gellynck, X. (2013). Operational performance and critical success factors of lean manufacturing in European food processing SMEs. *Trends in Food Science & Technology* 31(2): 156–164.

Ghobadian, A. and Gallear, D. N. (1996). Total quality management in SMEs. *OMEGA* 24(2): 83–106.

Juran, J. M., Bigliazzi, M., Mirandola, R., Spaans, C. and Dunuad, M. (1995). A history of managing for quality. *Quality Progress* 28(8): 125–129.

Lee, C. Y. (2004). TQM in small manufacturers: An exploratory study in China. *International Journal of Quality and Reliability Management* 21(3): 175–197.

Oakland, J. (2000). *TQM Text With Cases.* 2nd ed. Oxford: Butterworth-Heinemann.

Shea, J. and Gobeli, D. (1995). TQM: The experiences of ten small businesses. *Business Horizons* 38(1): 71–77.

Snee, R. D. and Hoerl, R. (2003). *Leading Six Sigma.* Upper Saddle River, NJ: Prentice-Hall.

Wessel, G. and Burcher, P. (2004). Six Sigma for small and medium-sized enterprises. *TQM Magazine* 16(4): 264–272.

chapter three

Lean Six Sigma

3.1   What is Lean production system?

The manufacturing system has witnessed a transformation of craft production to mass production and Lean production. The key variables governing this transition include product complexity and market dynamism. Mass production fulfils economies of scale, i.e. unit cost reduces as a result of high volume production. The modern customer expects a product to be available at an optimum cost and when it is required. To attract the customer and to stay competitive in the market, organisations should constantly explore ways to reduce cost and improve productivity. One of the best avenues to improve the performance of the company is the implementation of Lean production systems. Ohno (1988) stated that the fundamental goal of the Toyota production system (1988) is to eliminate waste; this is achieved by quality control and quality assurance. Ohno (1988) emphasised the production of only the required type of units at the required time and in the required quantities. This enabled Toyota to become market leader in the worldwide automotive industry. The term Lean manufacturing was coined in the 1990s when a book titled *The Machine That Changed the World* was written by James Womack et al. (1990). This book combined production methods practiced in the US, Europe and Japan and referred to in the publication as 'Lean manufacturing'. Then, many production engineers and experts attempted to make production methods more efficient. Lean manufacturing enables streamlining the production system to achieve cost savings and customer satisfaction and further profit improvement.

The term Lean from the manufacturing viewpoint denotes identification and elimination of wastes in involved processes (Womack et al. 1990). Liker and Wu (2000) defined 'Lean' as 'a philosophy of manufacturing that is based on developing the highest-quality products with lowest cost and delivered on time'. A Lean production system generates wider product variants, at low cost, with higher productivity levels and delivery speed at optimum quality. Lean production techniques facilitate improvements in efficiency, response speed and production flexibility, which are realised by industrial enterprises such as Toyota.

Lean manufacturing is a business philosophy that facilitates continuous improvement of processes involved in the manufacturing scenario, independent of what types of products are being manufactured (Shah and Ward 2007). Lean manufacturing includes elements, rules and tools that are based on value creation (Gopalakrishnan 2010). Lean production focuses on minimising wasteful process steps and improving the speed of manufacture. Lean principles ensure an established track record of operational and strategic success, which facilitates enhanced customer value.

Every product has a value. A product or service is an output of a process that includes a series of steps. Value addition happens at all stages of the process. Product value implies customer willingness to pay. The customer requires a product with maximum value for low cost. Waste implies an activity for which the customer is not willing to pay. Waste is an entity that consumes resources or time but does not enhance product value.

3.2   Key principles of Lean production system

There are certain key principles of Lean manufacturing which require clear understanding to effectively deploy Lean (Gopalakrishnan 2010). Lack of understanding of these principles will end in failure due to lack of commitment. Without commitment, the process becomes ineffective.

Five principles of Lean manufacturing (Gopalakrishnan 2010; Dahlgaard and Dahlgaard-Park 2006):

1. Value from customer viewpoint
2. Value imbibed process sequence
3. Smooth flow of value stream
4. Pull production
5. Continuous improvement and sustenance

1. *Value from customer viewpoint*: Customer value denotes the customer's perceived preference for product characteristics evaluation to facilitate the achievement of customer goals. This definition implies customer and business perspectives. Customer value implies the difference between the values that the customer benefits from by acquiring and using a product and the cost of obtaining the product. Value denotes customer willingness to pay and its attached cost. The best price does not imply the lowest price but rather the best integration of quantity, quality and delivery. The product must induce in the customer the delight that it has value. To determine the product value, the process sequence from beginning to end needs to be analysed.

2. *Value imbibed process sequence*: Value stream denotes the sequence of all processes where cost is incurred. It is a sequence of activities needed to design, manufacture and develop a specific good or service, along which information, materials and worth flows. Value stream denotes product manufacture from the generation of an order until the product is delivered to the customer. Value stream

implies all essential steps to develop a product or service to the customer from start to end:

a   The process in value stream needs to be classified into activities adding value to the customer (value added), activities not adding value to the customer (non-value added) and activities needed for product completion (necessary but non-value added).

b.   The flow of value stream must be smooth, and obstacles in the value stream need to be eliminated.

Value-added activities typically add value from the viewpoint of the customer and denote customer willingness to pay. These activities enable the attainment of competitive performance. Non-value-added activities are typical waste, and they must be eliminated. Necessary but non-value-added activities are forcefully induced in the process because of government regulation.

3. *Value flow*: The product production using value stream based on customer requirements is called value flow. Flow must be smooth without any obstacles. The company's management must ensure that there are no bottlenecks in the process and the flow of value stream is smooth. Obstacles reduce the velocity of flow and in turn affect lead time and increase cost. Flow should also take care of inventory. The obstacles for smooth flow may be due to improper layout, improper tooling, downtime, equipment breakdown or large quantity of stock. Appropriate actions need to be effected to overcome the obstacles for flow velocity improvement.

4. *Customer pull*: In a push system, the product is manufactured without understanding the customer need and gets pushed to market so as to sell to the customer. A pull system focuses on producing products purely based on customer requirement. A pull system minimises and further eliminates inventory. Pull systems denote a response to customer demand rate, i.e. actual customer demand that enables the supply chain.

5. *Continuous improvement and sustenance*: The attainment of the organisation's goal depends on a continuous improvement mindset. The term 'continuous improvement' implies incremental improvement of products, processes or services over time, with the objective of waste reduction to improve workplace functionality, customer service or product performance. It is necessary to sustain the process, procedures, standards and efficiency and ensure that the organisation doesn't deviate from set goals. Perfection could be attained only by bringing about involvement from personnel involved in the process execution. Long-term management policy needs to be deployed for implementing Lean enterprise.

3.3   Benefits of Lean production system

The benefits from the implementation of Lean production are presented as follows (Dennis 2007):

1. *Improved quality*: Lean facilitates quality improvement through the application of problem-solving techniques and mistake-proofing mechanisms. Problem-solving techniques overcome the root cause of the problem, and mistake proofing prevents the recurrence of the problem, thereby facilitating quality improvement.
2. *Improved visual management*: Lean production enables visual management by incorporating a visual scan to recognise abnormalities. Lean production incorporates visual control mechanisms to enable ease of detection of abnormalities and streamlined processes.
3. *Increased efficiency*: Lean production incorporates line balancing and standardised work to ensure effective working and improvement. Standardised work procedures ensure uniform productivity.
4. *Ease of team management*: Lean production facilitates easy management of teams by proper work instructions and standardised work with analysis of problem areas.
5. *Total company involvement*: Lean production is a management philosophy that promotes involvement and teamwork across the entire organisation. Team culture enables the attainment of Lean benefits.
6. *Problem elimination*: Lean production systematically analyses and investigates the issues to determine root causes to overcome the problems. Also, actions will be initiated for vital few causes.
7. *Reduced space*: Lean initiatives reduce inventory and other forms of waste to effectively utilise the space. The waste reduction initiatives facilitate space creation.
8. *Safer work environment*: Lean production incorporates visual management and 5S* to ensure organised and safe workplaces. Workplace organisation facilitates streamlined processes and other associated benefits.
9. *Improved employee morale*: On deployment of Lean initiatives, employee morale could be realised through employee involvement and empowerment as part of the team. Rewarding schemes for employees on Lean improvement suggestions will be institutionalised to improve their motivation.
10. *Reduced lead time*: Lean facilitates faster execution of a business process with fewer delays so as to facilitate the timely launch of the

---

* 5S is a methodical way to organise your workplace and your working practices. 5S stands for Seiri (sort), Seiton (straighten), Seiso (shine), Seiketsu (standardise) and Shitsuke (sustain).

product in the market. Process cycle time and changeovers will be minimised to reduce lead time.

11. *Improved customer service*: Lean enhances customer service by delivering what precisely the customer wants and at the time they require. Lean ensures production in pace with demand so as to enable organisations to operate with minimal inventory.

12. *Increased responsiveness*: Business process from a Lean perspective will be faster, and each process is linked to organisational supply chains. It generates financial benefits to the organisation.

13. *Improved office performance*: Lean facilitates reduced order processing errors, streamlined customer service functions and reduced paperwork. Automation of office tasks will be facilitated by implementing Lean initiatives.

14. *Improved productivity and profitability*: Lean initiatives ensure productivity and performance by improving bottom-line benefits through reduction of production costs.

3.4   What is Six Sigma?

Motorola was facing extreme pressures from overseas competition, particularly Japan. While it is impossible to set a definitive date for the beginning of Six Sigma, around 1987 Bill Smith (a reliability engineer) and others began improvement projects that in many ways looked similar to Total Quality Management (TQM) projects (Harry and Schroeder 2000). Eventually, Mikel Harry and others helped Smith formulate this approach into an overall business initiative aimed at protecting Motorola's pager business (Pande et al. 2000). They named the initiative 'Six Sigma' based on the desire to reduce variation to the extent that specification limits for key process metrics were six standard deviations away from target (Harry and Schroeder 2000).

Six Sigma has at least three meanings, depending on the context. First, it can be viewed as a measure of quality. Sigma is a Greek letter which measures the variation in a process. Achieving a Six Sigma measure of quality means that processes are producing less than four defects per million opportunities. Secondly, Six Sigma can be viewed as a business improvement strategy and a philosophy. Thirdly, it is a problem-solving methodology that seeks to find and eliminate the causes of defects or mistakes in business processes by focusing on process outputs which are critical in the eyes of customers. The statistically based problem-solving methodology of Six Sigma delivers data to drive solutions, delivering dramatic bottom-line results (Snee and Hoerl 2007).

Motorola achieved tangible results, and other organisations began to take notice. Honeywell and Allied Signal, other organisations in similar markets to Motorola, launched Six Sigma initiatives around 1990. These also met with success. However, it was when General Electric (GE) CEO Jack Welch loudly proclaimed that GE was jumping into the Six Sigma game in late 1995 that the initiative moved off the back pages of the business section to the front page of the newspaper.

Today, Six Sigma has been widely accepted by a large number of manufacturing companies as well as service organisations, in particular, financial services such as Bank of America, J P Morgan Chase, Citibank and Bank of Montreal and health services such as Commonwealth Health Corporation and Mount Carmel Health. A number of public sector organisations have also embraced Six Sigma as a strategy to achieve process excellence and consistency in the delivery of services to customers.

3.5   Some common myths of Six Sigma

There is a pervasive perplexity and misinterpretation of what 'Six Sigma' is about. Here are just a few of the myths and the truths.

3.5.1   Six Sigma is another management fad

Of course, Six Sigma could be fad in an organisation if the leaders of the organisation treat it as a fad – 'something we do because it is popular and others are doing it'. Six Sigma is not, however, a fad in most of the organisations I am aware of; certainly, it is not a fad in those organisations that do the deployment properly (Antony 2007). To quote a Honeywell manager, 'Six Sigma works if you follow the process. If it is not working, you are not simply following the process'.

Six Sigma as a management fad that has lasted for nearly 30 years in some world-class corporations seems to be an oxymoron. Six Sigma has clearly added significant bottom-line value to organisations that have implemented it seriously. Conversely, it must be admitted that many organisations have blindly jumped on the Six Sigma bandwagon without carefully considering the level of commitment required. Such companies have either failed or had minimal success. For these organisations, Six Sigma was simply a fad, the flavour of the month (Antony 2007).

3.5.2   Six Sigma is all about statistics

There is another common perception that Six Sigma focuses on only training in various statistical tools and techniques and almost ignores the human factor (building of company culture by everybody's involvement and commitment for continuous improvement). The statistical terminology 'sigma' provides an impression of Six Sigma being a statistics and measurement program. Six Sigma is not just about statistics. The Six Sigma drive for defect reduction, process improvement and customer satisfaction is based on the 'statistical thinking' paradigm, a philosophy of action and learning based on process, variation and data. Statistical thinking is fundamental to the methodology because Six Sigma is action oriented, focusing on processes used to serve customers and defect reduction through variation reduction and improvement goals (Snee 2004).

3.5.3   *Six Sigma works only in manufacturing settings*

The relevance of Six Sigma extends beyond manufacturing to services, government and public sector, healthcare and non-profit organisations. Motorola developed Six Sigma and implemented it first in manufacturing. From 1990 onwards, they started implementing it in the non-manufacturing areas of the company. It was reported at the European Quality Forum in Berlin that Motorola managed to save $5.4 billion in non-manufacturing processes from 1990 to 1995.

Six Sigma offers a disciplined approach to improve service effectiveness (i.e. meeting the desirable attributes of a service) and service efficiency (i.e. time and costs). The objective of a Six Sigma strategy in service processes is to understand how defects occur and then to devise process improvements to reduce the occurrence of such defects, which improves the overall customer experience and thereby enhances customer satisfaction (Antony 2004). Experts agree that the most common reason that service-oriented organisations stay away from Six Sigma is that they see it as a manufacturing solution. One of the major hurdles that service-oriented organisations must overcome is the notion that, because their company is human driven, there are no defects to measure. This is a pure misconception according to various practitioners and experts of Six Sigma.

3.5.4   *Six Sigma works only in large organisations*

Six Sigma is about problem solving, and problems are everywhere. It does not matter what type or size of business this problem-solving methodology is applied to. You might be a wholesaler, a retailer, a manufacturer or a service organisation. No matter whether it is a 300-employee company or a 10-employee family business, Six Sigma will work as long as you follow the process effectively (Brue 2006).

Snee and Hoerl (2003) argue that there is nothing inherent in Six Sigma that makes it more suitable for large companies. They also suggest that the greatest barrier to implementation in small companies to date has been the way the major Six Sigma training providers have structured their offerings. Six Sigma has evolved into a business strategy in many organisations today, and its importance in small and medium sized enterprises (SMEs) will continue to grow every day because of the growing significance of supply chain issues.

3.5.5   Six Sigma is the same as Total Quality Management

Deming, one of the quality gurus of the 20th century, argued that TQM is terminologically vague, stating 'the trouble with total quality management, the failure of TQM, you can call it, is that there is no such thing. It is a buzzword. I have never used the term, as it carries no meaning' (Deming 1994).

There are three aspects of the Six Sigma strategy that are not emphasised in TQM. First of all, Six Sigma is result oriented and therefore places a clear focus on the bottom-line business impact in hard dollar savings. No Six Sigma project will be approved unless the team determines the savings generated from it. Second, the Six Sigma methodology define-measure-analyse-improve-control (DMAIC) links the tools and techniques in a sequential manner. Finally, Six Sigma creates a powerful infrastructure for training champions, master black belts, black belts, green belts and yellow belts (Snee 2004; Antony et al. 2005, Pande et al. 2000; Adams et al. 2003).

3.6   An overview of Six Sigma methodology

In this section, the authors would like to give readers a quick overview of the Six Sigma problem-solving methodology. However, a detailed explanation of the methodology will be provided in Chapter 6 of the book. Six Sigma utilises a powerful five-stage data-driven methodology to improve processes. The five stages of the Six Sigma methodology are:

Define: In this stage, one has to define the problem and the process the problem is associated with. The project goals and milestones will be decided, and customer requirements (internal and external) should be defined.

Measure: In this stage, one has to measure the baseline performance of the process under study. The main purpose of this stage is to collect valid and reliable data pertinent to the scope of the project.

Analyse: In this stage, one has to determine the root causes of poor performance or excessive variation which lead to defects or errors in the process under study. A number of statistical tools can be used to analyse the data and determine the potential root causes of the problem.

Improve: In this stage, one has to develop potential solutions which can improve the process performance and reduce the impact of the problem at hand.

Control: The purpose of this stage is to sustain the improved performance, generate a detailed solution monitoring plan, observe implemented improvements for success, update plan records on a regular basis and maintain a workable employee training routine.

3.7   Benefits of Six Sigma

Six Sigma enables organisations to improve their processes, making them more capable of delivering what the customer wants right first time. Those organisations that implement Six Sigma correctly achieve significant benefits that contribute to competitive advantage and to changing the culture in an organisation from reactive problem solving to proactive problem prevention. The following are some of the potential benefits of Six Sigma:

- *Increased revenue*: Six Sigma increases revenue by enabling your organisation to do more with less (i.e. you should be able to produce more products or deliver more services with less resources).
- *Reduced operational costs*: Six Sigma reduces costs associated with scrap, rework, repair, replacement, warranty, downtime, etc.
- *Improved employee morale*: Six Sigma can improve the morale of employees by involving them in the improvement process. It develops a sense of ownership and accountability for your employees.
- *Reduced fire-fighting*: Effective use of Six Sigma can reduce the costs associated with misdirected problem-solving efforts or fire-fighting.
- *Improved problem-solving skills*: Six Sigma utilises a set of tools within the problem-solving methodology, and people who are involved with Six Sigma project will get an opportunity to learn how the tools work in solving real-world problems.
- *Improved communication*: As Six Sigma demands teamwork, the communication between team members can be improved, and moreover the communication from team members to senior management team would also improve significantly from various interventions and through regular management review meetings of projects.
- *Increased quality and reliability*: Six Sigma methodology can be used to reduce defect rates and even prevent defects from occurring in processes. This would lead to increased product quality and reliability.

3.8   Some pros and cons of Lean and Six Sigma

3.8.1   Some pros of Lean

The following are some of the pros of a Lean production system (Schonberger 2008):

- *Positive workforce effects*: Lean strategies are often based around worker initiatives, assuming that those doing the work are the best source for ideas about improving the way it is done. Workers are more apt to feel involved and satisfied about themselves, their workplace and the work they do.

- *Conducive work environment*: With Lean manufacturing, cultures become standardised, and unwanted behaviours of employees and management are for the most part gone. This creates a more pleasant work environment.
- *Reduced floor space required*: Experts have estimated that if Lean manufacturing techniques are adopted correctly, it will help companies reduce the requirement for physical floor space by about 5%–30%.
- *Improved customer relationship management*: The Lean applications in an organisation can positively influence its relationship with its customers.
- *Change of attitude*: Implementing Lean production often demands a significant change in an organisation's attitude, which can be very challenging if an organisation is not well slated to deal with the changes.

3.8.2   Some cons of Lean

The following are some of the cons of a Lean production system (Lindlof and Soderberg 2011):

- *Lack of standard methodology*: There is no standard methodology to be followed by Lean practitioners in organisations for the deployment of Lean.
- *Over-focus on present*: Lean stifles creativity or experimentation, which not only hampers the organisation from responding to changes better, but also makes it difficult to realise sudden opportunities that have become the norm in a fast-changing external environment.
- *Not good for achieving process stability*: Lean does not help organisations to achieve stability for their critical processes and therefore process capability.
- *Not ideal for high-value low-volume settings*: Lean may not be the best methodology to use if you produce products which are of high value but of low volume.

3.8.3   Some pros of Six Sigma

The following are some of the pros of Six Sigma:

- Six Sigma as a business process improvement strategy places a clear focus on achieving measurable and quantifiable financial returns to the bottom line of an organisation.
- Six Sigma as a problem-solving methodology (DMAIC) utilises the tools and techniques for fixing problems in business processes in a

sequential and disciplined fashion. Each tool and technique within the Six Sigma methodology has a role to play and when, where, why and how these tools or techniques should be applied is the difference between success and failure of a Six Sigma project.

- Six Sigma creates an infrastructure of deployment champions, master black belts, black belts and green belts who lead, deploy and implement the approach.
- Six Sigma emphasises the importance of data and decision-making based on facts and data rather than assumptions and hunches! Six Sigma forces people to put measurements in place.

3.8.4   Some cons of Six Sigma

The following are some of the cons of Six Sigma:

- The challenge of having quality data available, especially in processes where no data are available to begin with (sometimes this task could take the largest proportion of the project time).
- The calculation of defect rates or error rates is based on the assumption of normality. The calculation of defect rates for non-normal situations is not yet properly addressed in the current literature of Six Sigma.
- The statistical definition of Six Sigma is 3.4 defects or failures per million opportunities. It would be illogical to assume that all defects are equally good when we calculate the sigma quality level of a process. For instance, a defect in a hospital could be a wrong admission procedure, lack of training required by a staff member, misbehaviour of staff members, unwillingness to help patients when they have specific queries, etc.
- Non-standardised procedures in the certification process of Six Sigma black belts and green belts is another limitation. This means not all black belts or green belts are equally capable. Research has shown that the skills and expertise developed by black belts are inconsistent across companies and are dependent to a great extent on the certifying body.
- The start-up cost for institutionalising Six Sigma into a corporate culture can be a significant investment. This particular feature would discourage many SMEs from the introduction, development and implementation of a Six Sigma business process improvement strategy.
- Six Sigma can easily digress into a bureaucratic exercise if the focus is on such things as the number of trained black belts and green belts, number of projects completed, etc., instead of bottom-line savings.

3.9 Why Lean Six Sigma?

Deploying Six Sigma in isolation cannot remove all types of waste from the business process, and deploying Lean management in isolation cannot bring a process into the state of statistical control and remove variation from the process (Corbett 2011). Therefore, some companies have decided to merge both methodologies to overcome the weaknesses of these two continuous improvement (CI) methodologies when they have been implemented in isolation and to come up with a more powerful strategy for CI and optimising processes (Bhuiyan et al. 2006).

The integration of these two approaches gives the organisation more efficiency and effectiveness and helps to achieve superior performance faster than the implementation of each approach in isolation (Salah et al. 2010). The popularisation and the first integration of Lean Six Sigma (LSS) were in the US by the George Group. However, the term LSS was first introduced into the literature in 2002 as part of the evolution of Six Sigma (Timans et al. 2012). Since that time, there has been a noticeable increase in LSS popularity and deployment in the industrial world especially in large organisations in the West such as Motorola, Honeywell, GE and many others (Timans et al. 2012; Laureani and Antony 2012) and in some small and medium sized manufacturing enterprises (Kumar et al. 2006).

LSS is a business strategy and methodology that increases process performance resulting in enhanced customer satisfaction and improved bottom-line results ($). It is also being widely recognised that LSS is an effective leadership development tool. Leaders enable an organisation to move from one paradigm to another; from one way of working to another way of working. In making these shifts, work processes of all kinds get changed. LSS provides the concepts, methods and tools for changing processes. LSS is thus an effective leadership development tool in that it prepares leaders for their role, leading change.

According to Arnheiter and Maleyeff (2005), in a highly competitive environment, diminishing returns may result when either Lean or Six Sigma is implemented in isolation. They argue that an LSS organisation would include the following three primary tenets of Lean management:

1. It would incorporate an overriding philosophy that seeks to maximise the value-added content of all operations.
2. It would constantly evaluate all incentive systems in place to ensure that they result in global optimisation instead of local optimisation.
3. It would incorporate a management decision-making process that bases every decision on its relative impact on the customer.

There is no universal determinant of when to use Lean and when to use Six Sigma. Six Sigma and Lean concepts offer complementary tool sets which, together with each other and with other best management practices, offer a comprehensive means of transforming a business from operational chaos at one extreme to operational excellence at the other. Bertel (2003) highlighted the point that using either one of them alone has limitations: Six Sigma will eliminate defects in processes, but it will not address the question of how to optimise process flow. In contrast, Lean principles are not very helpful in achieving high capability and high stability processes. There are many advantages to using strategic Six Sigma principles in tandem with Lean strategy. Although Lean methodology can lead to quick process improvements, it is sometimes difficult for a company to leverage such improvements companywide because no infrastructure exists to do so quickly and efficiently (Sharma 2003).The strategic use of Six Sigma principles and practices ensures that process improvements generated in one area can be leveraged elsewhere to maximum advantage, resulting in a quantum increase in product quality, process performance or organisational performance.

3.10   Benefits of Lean Six Sigma

In order to understand the real benefits of LSS in various manufacturing settings, the authors reviewed 20 case studies published in the literature (Snee 2010). The type of case studies represent aircraft manufacturing, printed circuit board (PCB) manufacturing, tyre manufacturing, automobile accessories manufacturing, automotive valves manufacturing and semi conductor devices manufacturing to name but a few. Moreover, the reviewed case studies have been published in referred international journals with some good credibility. The case studies represent various countries such as the US, China, Malaysia, New Zealand, India, Taiwan and the Netherlands. A review of 20 case studies on LSS revealed the following top 10 benefits:

1. Increased profits and financial savings
2. Increased customer satisfaction
3. Reduced operational cost
4. Reduced cycle time
5. Improved key performance metrics
6. Reduced defects in processes
7. Reduced machine breakdown time
8. Reduced inventory
9. Improved quality
10. Increased production capacity

Other soft benefits such as increased employee morale towards creative thinking and a reduction in workplace accidents as a result of housekeeping procedures also appeared in a number of cases.

3.11   Challenges in the implementation of Lean Six Sigma

As with any change in how we work, there are challenges to the new way of working that must be overcome. No discussion of LSS execution is complete without addressing the hurdles that organisations face in implementing LSS and how to overcome them. Here are a few common roadblocks in successfully implementing LSS in an organisation, and how to eradicate them:

- *Lack of understanding of Lean Six Sigma methodology*: The first is resistance due to lack of understanding of LSS methodology and a lack of belief that it will work. All are susceptible to these challenges, management and others alike. Education can help, but successful projects are usually the best vehicle to reduce this concern. In the final analysis, it is a leap of faith that the approach will work in your organisation. Successful projects silence the doubters, increase the self-confidence of the proponents of the approach and justify the leap of faith (Snee 2010). Companies can overcome this obstacle by committing fully to the process and employing and supporting LSS experts to ensure that the company is deploying the methodology and not just using the terminology. These experts also keep the project focused on core operations where they can make the most difference, not just on the simple changes and the low-hanging fruit.
- *Lack of LSS deployment road map*: A properly constructed deployment plan will lay out the project selection process and how the initiative will be sustained over time. Project selection and sustaining the effort are arguably the two most difficult aspects of LSS deployment. Poor execution of LSS happens when process improvements are not aligned with the organisation's goals, when the project is based on reactively solving problems instead of meeting strategic objectives. In many organisations, projects lack effective leadership and are managed inefficiently. When leadership is committed to applying the Six Sigma methodology, assigns top talent to project teams, puts the project through a formal selection and review process and provides the required resources, the odds of LSS success increase dramatically.
- *Lack of sustainability*: Although a number of organisations embark on the LSS journey, very few are sustaining the initiative over a period of time. Although most agree that LSS is here to stay, they also agree that learning how to sustain the results seems problematic at best and unattainable at worst. Reverting to the old way of doing things is common if sustainability measures are not a part of

the methodology. The following tips might be useful for sustaining an LSS initiative in any industrial setting.

- Establish process ownership and make sure that people are accountable for their own processes across the business.
- Focus more on 'process improvements' instead of measuring the success of the initiative based on the number of people who have been trained.
- Make LSS a continuous improvement program for everyone in the business and encourage cross-functional problem solving to break down silos across various business functions.
- Make sure that employees are provided with the necessary tools for process improvement so that they can continuously improve their processes.
- Make celebrating success a priority and recognise and reward employees for the execution of projects which lead to significant process improvements.
- Establish visible and committed leadership to make sure that LSS is the core strategy for process improvement in the organisation and the senior management team must be involved in the selection of strategic LSS projects which are aligned with the corporate goals.
- Cultivate an organisation of learning and innovation.

3.12 Summary

This chapter provides readers with an introduction to the Lean production system, Six Sigma and the rationale behind the integration of two of the most powerful process excellence methodologies we have ever witnessed in modern organisations. The chapter presents the benefits of both Lean and Six Sigma methodologies, some of the fundamental challenges in the deployment of LSS, and the benefits from the implementation of Lean, Six Sigma and LSS in manufacturing settings. The chapter also highlights some of the strengths and limitations of both Lean and Six Sigma methodologies.

References

Adams, C., Gupta, P. and Wilson, C. (2003). *Six Sigma Deployment*. Burlington, MA: Butterworth-Heinemann.

Antony, J. (2004). Six Sigma in the UK service organisations: Results from a pilot survey. *Managerial Auditing Journal* 19(8): 1006–1013.

Antony, J. (2007). Is six sigma a management fad or fact. *Assembly Automation* 27(1):17–19.

Antony, J., Kumar, M. and Madu, C. N. (2005). Six Sigma in small and medium sized UK manufacturing enterprises: Some empirical observations. *International Journal of Quality & Reliability Management* 22(8): 860–874.

Arnheiter, D. and Maleyeff, J. (2005). The integration of lean management and Six Sigma. *The TQM Magazine* 17(1): 5–18.

Bertel, T. (2003). Integrating Lean and Six Sigma – the power of an integrated roadmap, www.isixsigma.com (Retrieved 21 July 2015).

Bhuiyan, N., Baghel, A. and Wilson, J. (2006). A sustainable continuous improvement methodology at an aerospace company. *International Journal of Productivity and Performance Management* 55(8): 671–687.

Brue, G. (2006). *Six Sigma for Small Business*. Madison, WI: CWL Publishing Enterprises.

Corbett, L. M. (2011). Lean Six Sigma: The contribution to business excellence. *International Journal of Lean Six Sigma* 2(2): 118–131.

Dahlgaard, J. J. and Dahlgaard-Park, S. (2006). Lean production, Six Sigma quality, TQM and company culture. *The TQM Magazine* 18(3): 263–281.

Deming, W. E. (1994). Report card on TQM. *Management Review* 26: 7.

Dennis, P. (2007). *Lean Production Simplified: A Plain-Language Guide to the World's Most Powerful Production System*. Productivity Press.

Gopalakrishnan, N. (2010). *Simplified Lean Manufacture*. New Delhi: PHI Learning.

Harry, M. and Schroeder, R. (2000). *Six Sigma: The Breakthrough Management Strategy Revolutionizing the World's Top Corporations*. New York: Currency, http://www.venturehaus.com (Retrieved 16 August 2015).

Kumar, M., Antony, J., Singh, R. K., Tiwari, M. K. and Perry, D. (2006). Implementing the Lean Six Sigma framework in an Indian SME: A case study. *Production Planning & Control* 17(4): 407–423.

Laureani, A. and Antony, J. (2012). Standards for Lean Six Sigma certification. *International Journal of Productivity and Performance Management* 61(1): 110–120.

Liker, J. and Wu, Y. (2000). Japanese automakers US suppliers and supply-chain superiority. *Sloan Management Review* 42: 81–93.

Lindlof, L. and Soderberg, B. (2011). Pros and cons of lean visual planning: Experiences from four product development organisations. *International Journal of Technology Intelligence and Planning* 7(3): 269–279.

Ohno, T. (1988). *Toyota Production System: Beyond Large Scale Production*. Cambridge, MA: Productivity Press.

Pande, P., Neuman, R. P. and Cavanagh, R. R. (2000). *The Six Sigma Way: How GE, Motorola, and Other Top Companies are Honing their Performance*. New York: McGraw-Hill Professional.

Salah, S., Rahim, A. and Carretero, J. (2010). The integration of Six Sigma and Lean management. *International Journal of Lean Six Sigma* 1(3): 249–274.

Schonberger, R. J. (2008). *Best Practices in Lean Six Sigma Process Improvement*. Hoboken, NJ: Wiley.

Shah, R. and Ward, P. T. (2007). Defining and developing measures of lean production. *Journal of Operations Management* 25: 785–805.

Sharma, U. (2003). Implementing Lean principles with the Six Sigma advantage: How a battery company realised significant improvements. *Journal of Organisational Excellence,* Summer 22(3): 43–52.

Snee, R. D. (2004). Six Sigma: The evolution of 100 years of business improvement methodology. *International Journal of Six Sigma and Competitive Advantage* 1(1):4–20.

Snee, R. D. (2010). Lean Six Sigma: Getting better all the time. *International Journal of Lean Six Sigma* 1(1): 9–29.

Snee, R. D. and Hoerl, R. (2007). Integrating lean and six sigma: A holistic approach. *Six Sigma Forum Magazine* 6(3): 15–21.

Snee, R. D. and Hoerl, R. W. (2003). *Leading Six Sigma: A Step by Step Guide Based on Experience at GE and Other Six Sigma Companies*. Englewood Cliffs, NJ: FT Prentice-Hall.

Timans, W., Antony, J., Ahaus, K. and Solingen, R. (2012). Implementation of Lean Six Sigma in small and medium-sized manufacturing enterprises in the Netherlands. *Journal of Operational Research Society* 63(3): 339–353.

Womack, J. P., Jones, D. T. and Roos, D. (1990). *The Machine that Changed the World*. New York: Harper Perennial

chapter four

Lean Six Sigma road map for SMEs

This chapter discusses a general road map for the implementation of Lean Six Sigma (LSS) in small and medium sized enterprises (SMEs). When any new concept is introduced in an organisation, one needs to verify the readiness of the organisation to accept the implementation and the infrastructure requirements. Hence, this chapter incorporates sections on readiness factors (RFs) and infrastructure requirements along with an LSS road map for SMEs. All these sections of this chapter together act as a systematic approach for LSS implementation in any organisation. Thus, this chapter guides the people in SMEs to understand the general approach and requirements for LSS implementation.

4.1 Readiness factors for the successful introduction of LSS

The implementation of LSS requires a significant contribution from all levels of the organisation, and it aims at cultural transformation of the organisation towards process improvement initiatives. If the organisation is not ready for implementing LSS, it can lead to failure of the implementation initiatives, frustration among the workforce and resistance within the organisation towards any new initiatives in the future. Before an organisation decides to invest its resources in any LSS initiative, RFs may be taken into account, which are vital constituents that can make or break the chance of success (Antony 2014). Following are the commonly identified RFs and the associated variables for LSS implementation in SMEs.

4.1.1 RF1: Senior management commitment and involvement

LSS is a top-down initiative. Senior management commitment and involvement is a vital factor for initiating LSS activities in any organisation. Management must allocate resources for the execution of improvement initiatives. The following variables are important for this RF:

- Appropriate communication of LSS deployment and associated benefits
- Demonstration of involvement within the organisation
- Empowerment of employees
- Transparent management style
- Allocation of appropriate resources for project tasks

The commitment from upper management during all stages of LSS implementation should be actively visible throughout the entire process. It starts from the project selection and team identification. The top management should ensure that only high priority projects are selected and that the best resources available to the company are used.

4.1.2   RF2: Visionary leadership and culture inculcation

In the modern industrial scenario, leaders are responsible for setting a clear vision for LSS implementation within the organisation. Leaders in the organisation should take the initiative of cultural transformation of the organisation (Lim et al. 2015). The variables pertaining to this factor are listed as follows:

- Assisting and encouraging workers
- Institutionalising reward and recognition system
- Enabling employees to concentrate on process-related improvements
- Empowering people to make decisions and create leaders for tomorrow
- Mentoring style of leaders
- Communicating improvements and success

In most of the organisations where full-time LSS resources are not available, the team will be executing the LSS activities in addition to their usual responsibilities in the organisation. Hence, the team needs to put in additional effort for the success of an LSS project. If the organisation does not establish proper reward and recognition schemes for successful teams, it is very difficult to maintain the momentum of the LSS initiative.

4.1.3   RF3: Customer focus

One of the key aspects of all modern business strategies is to enhance customer satisfaction. An unwavering focus on customers forms the key aspect of LSS initiatives. Avenues for the fulfilment of customer expectations need to be embedded within the LSS framework. The following variables are vital for this RF:

- Ensuring customers are the focal point, and obtaining customer feedback
- Initiatives for expanding customer base
- Anticipating future customer needs and expectations
- Linking customer focus to organisation's strategy and further deployment activities

Projects are to be selected after understanding the perception of the customer regarding products supplied or services delivered. If the projects are selected based only on the observed problems internally, the result of an LSS movement will not necessarily create an impact on the organisation or the customer.

4.1.4 RF4: Selecting the right people

The success of LSS deployment depends on the selection of the right task force members. An appropriate composition of team members is essential for ensuring the success of an LSS deployment initiative. Each LSS project would require a team leader and its members. It is essential to include the best people from the respective process into the team. The team leader should have leadership qualities to take people along with him or her. He or she should have the ability to understand data and how to analyse the data, and he or she should be confident in communicating with management and the project team. The team leader is responsible for the timely completion of the project with expected results. The team members are to be selected from the respective sub-processes. They should learn the necessary skills needed, assist in the collection of data and carry out other activities required of the project. Contribution in terms of process knowledge and expertise is also expected from the members. Overall, the team should act as change agents in the organisation and be a party to the cultural transformation in the organisation towards data-based decision-making. Variables pertaining to this RF are listed below:

- Clear understanding of LSS efforts by team members
- Right composition of members with representation from key stakeholders
- Knowledge of LSS methodology and tools by members
- Flexibility, transparency and learning aptitude by team members

4.1.5   RF5: Linkage of LSS deployment to organisation's business strategies

The benefits of LSS deployment could be sustained if LSS initiatives are in alignment with the organisation's business priorities. LSS projects need to be planned in such a manner that they are aligned with the goals of the organisation and are related to critical processes or business issues. The variables pertaining to this RF are presented as follows:

- Alignment of LSS projects with the organisation's strategy, goals and objectives
- Focus on small-scale projects (also called quick-hit or low-hanging fruit projects)
- Clear understanding of LSS strategies within the organisation

During the LSS project selection stage, it needs to be ensured by the top management that all projects selected have a clear linkage with business objectives and results. This can be ensured by selecting projects from critical business and customer issues. Management should also prepare a mapping of project goals to company objectives, so that everyone understands the linkage of the LSS project with company goals. With this linkage, the organisation can ensure that the successful completion of the project will lead to improved business results. This also can ensure the support of all stakeholders for the LSS project.

4.1.6   RF6: Competence to develop effective framework

The success of the LSS framework depends on its effective integration of Lean and Six Sigma tools. The workforce must have a clear understanding of the LSS framework and must be trained on the deployment stages of the framework. Variables associated with this RF are presented as follows:

- Identification of appropriate Lean and Six Sigma tools
- Knitting of Lean tools within Six Sigma framework
- Training sessions to enable employees to understand the framework

4.1.7   RF7: Appropriate selection and usage of LSS metrics

LSS metrics must be appropriately selected and subjected to usage. Team members must be trained on the computation of LSS metrics. The organisations are benchmarked based on LSS metrics like sigma quality level, defects per million opportunities (DPMO), rolled throughput yield, etc. More discussion on the LSS metrics is provided in Chapter 5. Variables pertaining to this RF are as follows:

- Identification of appropriate LSS metrics
- Accurate computation of metrics

While computing the metrics, care must be exercised by the team to ensure that they utilise only those metrics which are appropriate for their respective processes. The general guidance for calculating these metrics is also provided in Chapter 5. Usage of too many metrics may complicate the entire scenario and generate confusion among the team members. Hence, during the training sessions, emphasis should be given to the selection of the right metric suitable for the right situation.

4.1.8   RF8: Education and training

Training is to be provided in the define-measure-analyse-improve-control (DMAIC) methodology to the people associated with the LSS implementation. The training should cover specific information required to run an LSS project, such as data-based decision-making and usage of statistical software to analyse data, and motivate people involved in processes to participate in the cultural changes in the organisation. The topics listed in Chapter 6 are to be addressed during the training. Expert involvement in training programs is essential. These experts can be people from inside or outside the organisation, who have knowledge on LSS methodology and have experience in handling LSS projects. The success of LSS programs depends on effective education and training within the organisation. The associated variables of this RF are presented as follows:

- Identification of training needs
- Formulation of training curriculum for different levels of executives and employees
- Periodic visits to showcase best LSS practices
- Expert involvement in handholding of projects during LSS deployment phases

Training must be planned for champions, black belts (BBs), green belts (GBs) and yellow belts (YBs) in LSS methodology. The duration of training for champions can be for 1 day, and for BB range from 10 to 14 days. The GB training duration is 4–5 days, and YB is 2 days. It is a good idea to provide awareness training for half a day or for 1 day (generally known as white belts – WBs) on LSS for all the people working in the organisation. This will help the workforce to understand the overall structure of the LSS implementation framework. Also, guidance needs to be provided to the team at different stages of project execution, and periodic review by the champion or management is essential to ensure that the projects are on track.

4.2   Lean Six Sigma implementation infrastructure

The provision of necessary infrastructure plays a very critical role in the successful implementation of LSS in any organisation. The resources include budgetary support for various activities, competent manpower, logistic facilities and other software and computational facilities.

The most critical among the resources is trained manpower. As discussed in the previous chapters, stakeholders in LSS implementation include champions, master black belts (MBBs), BBs, GBs and YBs. In a typical SME set-up, taking the services of full-time or even part-time MBBs and BBs may not be feasible due to various limitations including budgetary constraints. Most of the time, the projects need to be managed with GBs and YBs. A typical GB project can be of 3- to 4-month duration, and a YB project can be less than 2 months' duration. The expected saving from a GB project is approximately US$10,000 and a YB project is around US$1,000–US$2,000. It is also suggested to have two projects to be completed for GB certification, so that the GBs get good expertise in LSS concepts.

One of the important decisions that needs to be made at this stage is the selection of GBs and YBs. Management should ensure that the best possible resources in the organisation are selected as GBs and YBs so that the LSS projects can be successfully completed. People with proven track records with leadership skills should be selected to lead the projects. As LSS is new to the organisation, the people need to be trained as GBs and YBs. If the organisation can afford it, one person can be trained by a consulting/training firm as a BB and in turn, the BB takes care of the training of GBs and YBs internally. Apart from the training, the BB can provide guidance to the project team for successful completion of the project. This includes support in conducting appropriate analysis using statistical software at various phases of the LSS project.

It is recommended to select a maximum of four or five projects during the first year of the introduction of LSS in the organisation. Once the initial projects are successfully completed, the organisation and its employees will be confident in handling more and more projects. As a result, during the second year onwards, more projects can be executed.

The GB training can be of 4–5 days' duration in one or two phases. The YB training is typically for 2 days' duration. To bring general awareness in LSS among employees who are not covered under the GB/YB training in the organisation, a 1-day WB training session is recommended. This can help people understand the LSS methodology and ensure support for the initiative. Once the training is completed, the GBs or YBs can start working on the project. A project review is required periodically to ensure the smooth completion of the project successfully. A review after every phase is necessary to ensure that the respective phases are completed suitably. The progress in terms of the set time frame and results achieved should be reviewed by the champion so that the alignment of organisational priorities and project objectives is maintained (Breyfogle 2003).

Since LSS projects necessitate data-based decision-making, it is essential to have data collection and analysis at various stages of project execution. This requires the usage of statistical software and computers. Hence, one copy of Minitab or any other similar statistical software is required to be installed in the organisation. If it is difficult for the organisation to acquire a copy of the Minitab software, the initial statistical analysis can even be performed using Excel software. The GBs and YBs need to practice using the available software. The Minitab or Excel software can be used to carry out basic statistical analysis mainly during the measure and analyse phases of the LSS project. This includes different types of graphical analysis like Pareto, Histogram, Box Plot, etc., and statistical analysis like correlation and regression, tests of hypothesis, etc.

Management also should provide the necessary resources in terms of manpower or budgetary support to implement worthwhile ideas emerging out of the project. This forms the motivation and encouragement for the people in the organisation to take part in the LSS initiative.

4.3   A road map for implementing Lean Six Sigma

This section provides a general roadmap for the implementation of LSS in an organisation. The road map acts as an implementation framework/ guideline to enable SMEs to improve their process performance by LSS. The organisation-wide implementation of LSS is to be initiated in four phases. These phases include conceptualisation, initialisation, implementation and sustenance (Kumar et al. 2011). The proposed road map includes generic implementation steps for SMEs to deploy LSS. This road map helps the organisation to understand the steps to be followed for implementing the LSS projects in the organisation. As LSS is new to the organisation during the first year of implementation, this road map acts as a guideline for the teams. It is also suggested that teams can do further customisation of this road map after the first year of implementation of LSS in their respective organisations after incorporating the knowledge and experience gained from the projects. A conceptual road map for LSS implementation is presented in Figure 4.1.

4.3.1   Conceptualisation

This phase forms the preparatory stage wherein the foundation for LSS deployment will be initiated. LSS is implemented as a top-down approach;

Conceptualization	Initialization	Implementation	Sustenance

Identify the need of LSS	LSS project selection	Define phase
		
		• Project charter
		• SIPOC
		• CTQ Tree
			Control phase
			
			• Standardization of the solution
			  ▪ Control plan
			  ▪ Quality plan
			  ▪ Poka-Yoke
			  ▪ Flow chart

Resource planning	Selection of GBs, YBs	Measure phase
		
		• Data collection plan
		• MSA
		• Graphical summary
		• Stability study
		• Baseline status evaluation
			• Monitoring results
			  ▪ Control chart
			  ▪ Reaction plan

Management Buy-in	Team formation	Analyse phase • Process analysis • Cause and effect analysis • Data-based cause validation	• Horizontal deployment • Estimating the savings • Capturing the learning
	Training	Improve phase • Solution generation • Solution prioritization • Risk analysis • Piloting • Implementation	• Communicating success • Reward scheme

Figure 4.1 LSS implementation road map.

first, top management should evaluate the urgency for competitive advantage through implementation of LSS methodology. Top management should be convinced about the need for implementing LSS and should take the lead in communicating with employees in the organisation about the significance of LSS implementation.

LSS methodology can be employed in the following situations:

- There is a need to address critical business issues or customer requirements.
- Previous attempts to solve process problems through application of various methods were not successful.
- The organisation is looking at addressing the cost of poor quality–related issues in the organisation.

- The problems that need to be addressed are highly complicated, and there is no clue to the people in the organisation about the solution to these problems.
- Organisations are looking for long-term, sustainable solutions.

Budget and other resources required for the implementation of LSS also need to be planned at this stage. Lack of resources can lead to the failure of implementation of the LSS initiative.

4.3.2   Initialisation

This phase discusses the modalities for selecting the LSS project and identifying the resources required to execute the project. LSS projects are selected from a customer or business perspective. Those projects which can create an impact on the business or the customer are selected as LSS projects – hence the critical business processes and critical customer issues to be selected as projects. Once such problems are addressed, it can improve customer satisfaction and have huge financial benefits to the organisation. Thus, utmost care must be taken while selecting the project. The wrong project can lead to the failure of the entire LSS initiative in the organisation.

Another important aspect in LSS implementation is the formation of LSS project teams. In an SME set-up, the team typically consists of GBs and YBs. The GBs and YBs are to be selected from the best performers in the organisation. People with good analytical capability, innovative thinking and leadership qualities are ideal for these roles.

Management understands the priority of the organisation, and hence the involvement of top management in the selection of the projects is very important. All the selected projects have to be aligned with the management and business priorities.

4.3.3   Implementation

During the implementation phase, the LSS project is executed by the team by following the DMAIC methodology. Each phase of the DMAIC has specific activities to be completed. The define, measure, analyse and improve phases of the DMAIC focus on process improvement, whereas the control phase focuses on sustainability of the achieved results. Depending on the type and complexity of the problem to be addressed, different types of tools and techniques need to be applied for analysis and decision-making in each project. During the implementation stage, the define, measure, analyse and improve phases are executed in the project.

The objective of the define phase is to define a project with all the necessary details. This phase starts with preparation of a project charter for the selected project. Once the define phase is completed, it provides good clarity to the team about the project they are working with in terms of objectives, scope, defect definition, etc. This is a major step in the project execution stage. If this phase is not properly addressed, it can even lead to failure of the project.

During the measure phase, the baseline (current) performance level of the process is to be estimated. Hence, data are to be collected for all critical to quality characteristics (CTQs), and the baseline status is to be evaluated. The deliverable from the measure phase is the performance of the process calculated in terms of sigma rating of the CTQs.

The objective of the analyse phase in an LSS project is to study the process in detail to identify the root causes of the problem. Broadly there are two types of analysis performed during this phase. One is the 'process door' analysis and the other one is the 'data door' analysis. During the process door analysis, a detailed study of the existing process is performed. For the data door analysis, the data on various potential causes are gathered, and statistical analyses are performed. The deliverable in the analyse phase is the list of root causes identified. The identified root causes can be from either the process door analysis or the data door analysis.

During the improve phase, the solutions are identified for the selected root causes, and improvements are made in the process. Thus, at the end of the improve phase, solutions are implemented, and results are observed. The specific details of these phases are provided in Chapter 6 of this book.

4.3.4   Sustenance

The biggest challenge in any process improvement initiative is the sustainability of the achieved results. In SMEs, there can be quite a few challenges like a high attrition rate for employees, non availability of skilled manpower, lack of highly automated machinery, etc. Under this SME environment, once an LSS project is completed with good results, maintaining the same in the long run is extremely difficult.

Sustainability of the results can be achieved mainly through three important actions viz, standardisation, monitoring and training. For standardisation of the improved process, tools like control plan, quality plan, Poka-Yoke, flowchart of the improved process, etc., can be used. If the organisation is certified to the International Organisation for Standardisation (ISO), then it is a good practice to bring all the modified/newly introduced documents under the 'document control' system so that it can be ensured that everyone follows the improved process. Monitoring of achieved results is done through control charts for each CTQ. Depending on the type of data, one can use an individual chart,

X-bar–R chart, p-chart or u-chart for monitoring the process. A reaction plan can also be prepared to describe the actions to be initiated when something goes wrong.

A report on all the success stories of LSS implementation needs to be prepared by the team and communicated within the organisation. This can be communicated through internal presentations or publications in a newsletter or on the intranet, etc. Motivating employees by suitable reward schemes for successful projects is also very important in LSS implementation. This can create intrinsic motivation in employees to work on more and more projects. In the LSS methodology, the sustenance part is described in the control phase. Further details regarding the control phase are discussed in Chapter 6 of this book.

4.4 Managerial implications

A well-planned LSS implementation can lead to a rewarding experience and immense benefits for an organisation. On the other side, a failed implementation may lead to disappointing results – the failure of the entire implementation effort, and wastage of time and resources (Gijo 2011). The failure of the LSS initiatives can start from the project selection and team identification stage itself. If the right project is not selected with a competent team, the projects can be a failure. Lack of specific training of people and non-availability of other resources can also lead to failure. Due to high attrition rates in organisations, the sustainability of the achieved results will also suffer (Gijo and Rao 2005).

Successfully avoiding these common mistakes/shortcomings during LSS implementation will yield long-term benefits for an organisation and accelerate its march towards becoming the best in its class. The key to success in LSS is to identify these challenges early and take corrective actions to arrest the problems before they become an issue. This will help in timely implementation, necessary for deriving the benefits of LSS initiatives. With responsible professionals' conscious efforts to avert potential shortcomings, LSS has all the ingredients to continue to be successful for many years to come.

4.5 Summary

This chapter provides the readers with the RFs, the infrastructure requirements and a road map for successful implementation of LSS in SMEs. This provides a guideline for the organisation to understand the resources required to implement LSS. It also provides a step-by-step road map for project execution. The steps to be addressed in each phase of the LSS project deployment are explicitly listed. The steps towards sustainability of results, including standardisation of improved processes and monitoring of achieved results, are also discussed. Factors like management commitment, employee involvement and provision of resources including trained manpower are essential for the success of LSS deployment. Even the reward and recognition of successful teams is very critical for the LSS journey in the organisation. All these aspects of LSS implementation are discussed in this chapter. Thus, this chapter can act as a ready reference for the implementation of LSS in SME set-up.

References

Antony, J. (2014). Readiness factors for the Lean Six Sigma journey in the higher education sector. *International Journal of Productivity and Performance Management* 63(2): 257–264.

Breyfogle, III F. W. (2003). *Implementing Six Sigma: Smarter Solutions Using Statistical Methods*. New York: John Wiley.

Gijo, E. V. (2011). Eleven ways to sink your Six Sigma project. *Six Sigma Forum Magazine* 11(1): 27–29.

Gijo, E. V. and Rao, T. S. (2005). Six Sigma implementation: Hurdles and more hurdles. *Total Quality Management & Business Excellence* 16(6): 721–725.

Kumar, M., Antony, J. and Tiwari, M. K. (2011). Six Sigma implementation framework for SMEs: A road map to manage and sustain the change. *International Journal of Production Research* 49(18): 5449–5467.

Lim, S. A. H., Antony, J., Garza-Reyes, J. A. and Arshed, N. (2015). Towards a conceptual road map for statistical process control implementation in the food industry. *Trends in Food Science & Technology* 44: 117–129.

chapter five

Lean and Six Sigma metrics

5.1   Introduction

Lean Six Sigma (LSS) enables modern manufacturing/service small and medium enterprises (SMEs) to create and sustain competitive advantage. It has been observed from various organisations that the integrated LSS strategy facilitates organisations to improve quality, productivity and product and service consistency and reduce/eliminate different types of waste across the business. LSS integrates both waste reduction and value improvement tools from the Lean dimension and defect reduction and quality improvement tools from the Six Sigma dimension (Devadasan et al. 2012). In this context, appropriate Lean and Six Sigma metrics need to be assessed for performance improvement. Also, such metrics would serve as baseline data for enhancing the business process performance of modern organisations. LSS emphasises the measurement and evaluation of process performance results. The selection and usage of appropriate metrics in the context of LSS would enable organisations to identify and target the appropriate problems during LSS implementation projects, evaluate potential improvement initiatives and select action plans, establish baseline data for process enhancement, communicate the improvement results and continuously monitor the deployment of LSS initiatives. Lean metrics include *customer value, process flow, waste (8 forms), activities (value added, non-value added, necessary but non-value added), first-time quality, takt time, lead time, cycle time* and *changeover time.* Six Sigma metrics include *defects per million opportunities* (DPMO), *sigma quality level, throughput yield, rolled throughput yield, cost of poor quality* (COPQ) and *process capability indices* (PCIs). Also, the most common performance indicators for an LSS strategy are categorised based on time, cost, quality and output along with tangible bottom-line savings, profit margin, return on investment (ROI), market growth, revenue growth, employee morale and so on. This chapter presents the details of these metrics and provides illustrative examples for the same.

5.2   Introduction to common metrics of Lean

5.2.1   Value

Value is defined as 'any process or action for which the customer is willing to pay'. Toyota's 14 management principles define value as the quality of a product or service as defined by customers. Further, in the customer's view, value governs the transformation of a product or service and creates willingness to pay.

Womack and Jones term value as the capability provided to the customer at the right time at a price, as defined by the customer for a particular case. They also emphasise that value is the critical initiation for Lean thinking, and can only be defined by the ultimate end customer. They also argue that value is product oriented, and it is meaningful only when expressed in terms of a specific product.

Every product or service has a value. Value addition takes place at each and every stage. The value of raw material gets added at every stage and at the end, it becomes a product value. The value of a product or service is finalised by customer willingness to pay. The customer aspires for maximal value at a competitive price (Gopalakrishnan 2010).

5.2.2 Customer value

Every activity in a process incurs costs being paid for by the customer. The ultimate goal of a customer is value for money. A clear understanding of customer requirements is an essential part of any continuous improvement initiative. Thus, before commencing product or service development, product or service designers need to understand what product or service features add value to the customer. Waste refers to the entities that do not add value from the viewpoint of the customer and need to be reduced or even eliminated. To understand customer requirements clearly, a cross-functional team needs to be formed with executives from different departments to clearly understand what the customer expects. Value is attached to the cost of a product. It is recommended to work out the target cost by studying the process and eliminating visible waste. Also, the cost must not be compared with the competitor's price. It is highly important to eliminate non-value-added activities to bring down the cost. Value is generally perceived by the customer. A product or service adds value to the customer only when it fulfils the customer's known and perceived requirements.

5.2.3 Creating value

Lean principles emphasise value creation by specifying value from the customer's viewpoint, designing the product which can fulfil all issues desired by the customer, determining a value stream by mapping the processes, improving the value stream by enhancing its process capability, organising value-creating steps, creating flow with capable processes, creating pull of materials, parts and information and continuously improving the value chain.

5.2.4   Flow

- The movement of raw materials or information.

5.2.5   Value stream

- Refers to the sequence of the process from start to finish where the cost is incurred or value gets created.
- Indicates all the activities required to develop a product or service.
- The flow across the value stream must be smooth without any bottlenecks. The prevalence of bottlenecks in the value stream indicates the presence of waste and requires its removal. A smooth flow can be ensured with the involvement of all personnel associated with the process.

5.2.6   Value flow

- Refers to the products produced or services delivered using the value stream based on customer requirements.

5.2.7   Waste

- Waste in the context of Lean thinking implies any task or action for which the customer does not pay. Any entity in a value stream that incurs cost or time without value addition denotes waste.

Lean emphasises the ethos of continuous improvement, and the focus is to eliminate waste and non-value-adding activities across the core and supporting business processes in an organisation. Waste is defined as any operation or step or activity which does not add value (Tapping et al. 2002). Some waste reduction initiatives include stopping defective products at their source, combining processes that have similar attributes, eliminating unnecessary process tasks or procedures or steps and reducing the waiting time for materials, parts or people, to name but a few here.

According to the Toyota production system (TPS), wastes are classified into three different forms:

1. 'Mura' or waste due to variation or unevenness
2. 'Muri' or waste due to overburden
3. 'Muda' also known as waste

Further, muda (waste) is classified into two different categories, namely type-1 muda and type-2 muda:

1. *Type-1 muda*: These are activities (waste) that do not add value but are necessary for the system or process to function properly. This type of waste can be minimised but not necessarily eliminated completely.
2. *Type-2 muda*: These are activities (waste) that do not add value (waste) and are unnecessary for the system or process. This type of waste must be eliminated with immediate action.

Following are the various forms of waste that are most commonly associated with a Lean production system (LPS):

1. *Transportation*: Transportation is the movement of materials from one location to another, and this is considered as a form of waste as it does not any add value to the product. Excess transportation is caused by poor factory layouts, complex material handling systems, multiple storage and large batch sizes. This waste can be eliminated by designing a linear sequential flow from raw materials to finished goods, creating appropriate facility layout using facilities planning techniques, one-piece flow, signboards and colour lines.
2. *Inventory*: Material quantities including raw materials, work in process (WIP) and finished goods which go beyond the immediate need are considered as waste as they need space and incur cost for storing them. Inventory locks working capital and storage space. This form of waste can be eliminated by utilising a just in time (JIT) approach, one-piece flow, value stream mapping (VSM) and electronically controlled production using IT infrastructure and by eliminating buffer steps in production.
3. *Motion*: Movement of man or machine or material between work stations for the completion of a task is considered as another form of waste. It can be eliminated by following 5S practices, which creates a better organised workplace, and by labelling, sequencing boards, Kanban cards, material handling systems, standardised work techniques, designing an appropriate material handling system and arranging the machines in the manufacturing sequence to avoid unnecessary motion.
4. *Waiting*: The idle time generated when two interdependent processes are not absolutely synchronised. Waiting is considered as a waste as it disturbs the flow. This waste is caused by poor man/machine orientation, long changeovers and improper planning. It can be eliminated by following standardised work instructions, 5S, single minute exchange of dies (SMED), visual control and continuous flow principles.
5. *Overproduction*: A form of waste which occurs while producing a product before the customer requires it. Overproduction leads to high levels of inventory and increases storage cost. It is caused by

improper forecasting and inaccurate information on actual demand. It also leads to wastage of worker effort and unnecessary money locking. It can be eliminated by creating a pull-type manufacturing system, Kanban cards and VSM for analysing material and operations flows, by creating a proper balance between supply and demand and by using IT utilities.

6. *Over-processing*: Over-processing is the extra work that is performed beyond the standard required by the customer. It is caused by non-standardisation of processes and unclear information about the specifications of the product. It can be eliminated by conducting Kaizen activities by comparing customer requirements with manufacturing specifications, proper process design, Poka-Yoke (mistake proofing), smart processing and sequencing boards.

7. *Defects*: This refers to the quantity of production that is considered as scrap or requires reworking. Defects are caused due to inadequate training, operator errors and equipment malfunctioning. Defects can be reduced or eliminated by applying Poka-Yoke principles, conducting root cause analysis, practicing proper communication, creating work instructions as per standardised work and conducting effective training programs.

8. *Underutilisation of workforce expertise*: This is the eighth form of waste, which has been identified quite recently and is gaining in popularity in many service and public sector organisations. Modern organisations recognise that the workforce is one of their key assets. When the workforce ideas and creativity are not recognised, it becomes a waste for the organisation. Hence, this waste can be remedied by encouraging the workforce to suggest their ideas and encouraging workforce creativity. Table 5.1 presents examples for all categories of waste in an LPS.

5.2.8   Value-added activity

Value-adding activities denote activities that enhance product value and those activities for which the customer is willing to pay. They are also defined as activities that physically change the shape or characteristics of a product or assembly from its existing form. Any segment of the production or service process that enhances product value or service value from the customer viewpoint is a value-adding activity.

An activity must satisfy three criteria to become a value-adding activity:

1. Activity must transform the item towards completion.
2. Activity is done right for the first time without any rework.
3. Activity for which the customer is willing to pay.

Table 5.1 Examples for eight wastes

Waste category	Examples
Transportation	• Material transfer • Multiple handling
Inventory	• Excess safety stock • Excess inventory (raw material, WIP, finished goods) • Unsold items
Motion	• Unnecessary worker movement • Inappropriate layout
Waiting	• Slow pace • Excess queue • Late delivery • Material non availability
Overproduction	• Produce more than required • Delivery in advance
Over-processing	• Inappropriate processing steps • Excessive paperwork
Defects	• Out of specifications • Scrap • Rework
Underutilisation of workforce expertise	• Rejecting the workforce suggestions • Not encouraging workforce creativity

Source: Kavanagh, S. and Krings, D. (2011). *Government Finance Review*, 27(6–18).

5.2.9 *Non-value-added activity*

Non-value-adding activities denote activities that do not add value to the product or service. It denotes customer non-willingness to pay. Non-value-added activities increase the time involved in product manufacture but do not enhance the product value.

The organisation must take necessary efforts to eliminate non-value-adding activities from the system and can utilise the time saved in other improvement activities. Consider the example of inspection, which does not directly contribute to product value but is necessary until the process is incorporated with automated checking utilities. Table 5.2 presents examples of value-adding and non-value-adding activities.

5.2.10 *First-time quality*

A Lean metric evaluates the extent to which the parts are produced accurately the first time without any need for inspection or rework. It is one of the vital performance measures of a Lean manufacturing system.

Table 5.2 Examples for value-adding and non-value-adding activities

Value-adding activities	Non-value-adding activities
Direct machining	Reworking defective products
Assembling of parts	Product inspections
Preparing engineering drawings	Retesting of products
Raw materials	

5.2.11 Computation of first-time quality

A component consumes 1 minute to be produced. First-time quality for this part is 90% (or it can be said to have a 10% failure rate). Typical production day is one 8-hour shift = 480 minutes.

The available time of 430 minute/shift is obtained by subtracting the time for two 10-minute tea breaks and a 30-minute lunch break. Running this part on a single assembly line with a minute cycle time would result in 430 units being produced per shift. First-time quality of 90% actually produces 387 units per shift. This level of quality, or lack of quality, amounts to a loss of 43 minutes of available capacity.

5.2.12 Cycle time

- It is the actual time incurred to perform a task and proceed to the subsequent step. One of the major goals of Lean principles is to match cycle time with takt time.
- It is the time that elapses between one part coming out of the process and the next part entering it.

5.2.13 Takt time

- It is the rate at which the customer expects the organisation to perform product manufacture. Takt time is the ratio of number of working hours to number of orders per day.
- It is the time needed to complete an activity in order to meet customer demand.
- It is the rate at which an organisation must manufacture a product to fulfil customer demand. It implies the synchronisation of manufacture pace with sales speed. In order to synchronise between production and demand, takt time needs to be adjusted to increase or decrease order volume.

5.2.14 Lead time

- It refers to the time taken for one product to travel through the entire value stream from start to finish.

5.2.15 Changeover time

Changeover is the process of transforming a line or machine from producing one product to another. Changeover time is the time spent on conversion (includes switching fixtures, tools, programming and initial set-ups) of a line or machine to manufacture another product from the running product:

- It is the time spent modifying the setting from one product model to another.
- It is the actual time available in the shift for carrying out processes and changeovers.
- It is the time to transfer from manufacturing one product type to another.

5.2.16 Worked examples

5.2.16.1 Example 1

A manufacturing plant works two shifts of 8.5-hours basis with a lunch break of 30 minutes. The workers get 10 minutes at the start and end for cleaning purposes and 10 minutes for a tea break.

Compute takt time.
Total available time = 510 minutes
Available production time (AT) = 450 minutes
Customer demand (CD) = 200 parts/day
Takt time = AT/CD = 450/200 = 2.25 minutes

If the cycle time of all processes is closer to the takt time, the customer demand can be met without any delay.

5.2.16.2 Example 2

A raw material has to undergo three different processes (refer to Figure 5.1) to get converted into a finished product. C/T denotes the cycle time of process 1, process 2 and process 3.
Compute the lead time.

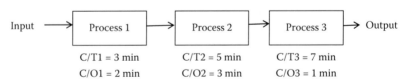

Figure 5.1 Lead time calculation.

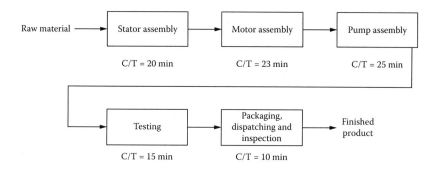

Figure 5.2 Lead time and takt time calculations.

5.2.16.2.1   Cycle time   Here, the cycle time for process 1 is 3 minutes. This implies that the total time required to complete process 1 is 3 minutes.

5.2.16.2.2   Lead time   In the above case, the lead time is 15 minutes (summation of cycle times of process 1, process 2 and process 3).

5.2.16.3   Example 3

A firm which manufactures submersible pumps has five stations (stator assembly, motor assembly, pump assembly, testing and packaging, dispatching and inspection). The cycle times of each station are described in Figure 5.2. The firm operates for 8 hours (480 minutes) per day with half an hour lunch break and two 10-minute tea breaks. The daily demand is 50 pumps.

From the given data,

Net available time = 480 − (30+10+10) = 430 minutes
Lead time = 20+23+25+15+10 = 93 minutes
Takt time = Net available time
Daily demand
= 430/50
= 8.6 minutes

This implies that the firm has to manufacture a pump every 8.6 minutes.

5.3   Introduction to common metrics of Six Sigma

The most commonly used metrics of Six Sigma are DPMO, sigma quality level, rolled throughput yield, CPQ and PCIs. For any critical to quality characteristics, these metrics can be calculated based on the data collected from the process.

5.3.1   Defects per million opportunities

The quality of output characteristics of any process is evaluated against a defined specification. During any such inspection process, each characteristic is checked against the possible opportunities for defects. Any non-conformity in a product or service is defined as a defect. Then, the defects per opportunities (DPO) can be evaluated by the formula:

$$DPO = \frac{\text{Total Defects}}{\text{Total Opportunities}}$$

Based on this, the DPMO can be obtained by the formula $DPMO = DPO \times 10^6$. This metric provides the defects out of a million opportunities.

5.3.1.1   Example 1

A vendor supplies 20,000 items, and 5% of these items are checked as per the sampling plan for five characteristics. During this inspection, 100 defects are observed. The DPMO of this process can be calculated as follows.

In this case, since the inspection was carried out for five characteristics, the opportunity for defect in a unit is five. As 5% of 20,000 items are inspected, the sample size is only 1,000 items.

The total opportunities per defect are $1000 \times 5 = 5000$. Then,
$DPO = 100/5000 = 0.02$
$DPMO = DPO \times 10^6 = 0.02 \times 10^6 = 20,000$

5.3.2   Sigma quality level

Sigma quality level or sigma rating is a commonly used metric in Six Sigma methodology. The sigma rating helps us to evaluate the performance of a process. This also helps us to compare performances of dissimilar processes, which is otherwise difficult.

When the inherent variability of the process is used to determine the sigma value (assuming mean is centred at the target), it is called SHORT-TERM sigma denoted by Z_{ST}. This is the best performance that the process is capable of achieving. Over a period of time, assignable causes creep in, and the capability of the process to meet the specified requirements diminishes. This sigma, which represents the capability of the process to meet the requirements over a period of time considering that extraneous conditions cause process shifts, is called the LONG-TERM sigma denoted by Z_{LT}. Normally, the short-term sigma is higher than the long-term sigma.

Table 5.3 Sigma rating and DPMO: A comparison	
Sigma rating	DPMO
1	691,462
2	308,537
3	66,807
4	6,210
5	233
6	3.4

If the shift is not specified, then $Z_{ST} = Z_{LT} + Z_{SHIFT}$. Generally in the long term, it is assumed that the Z_{SHIFT} is 1.5 and hence $Z_{ST} = Z_{LT} + 1.5$. The value of Z_{LT} can be obtained from the standard normal tables based on process data. The short-term sigma, Z_{ST}, is commonly known as sigma rating. A sigma rating of 6 indicates DPMO of 3.4. A comparison of the sigma rating and the corresponding DPMO values are presented in Table 5.3.

5.3.3 Rolled throughput yield

Rolled throughput yield is defined as the probability that a single unit can pass through a series of process steps free of defects. To calculate the rolled throughput yield for a multistage process, the first pass yield of sub-processes is to be calculated first. A product of the first pass yield of the individual processes gives the rolled throughput yield. For example, consider a process with four stages. The first stage starts with 100 units, and after processing 97 units were good. Similarly, after processing in the second and third stages, 96 units and 91 units, respectively, come out as good (Figure 5.3).

Then the rolled throughput yield = 97/100 × 96/97 × 91/96 = 0.91, which is less than the smallest yield value of the individual stages.

5.3.4 Cost of poor quality

COPQ is the cost of not doing things right first time. In other words, it is the cost associated with not meeting customer requirements the first time. COPQ has been widely used in many organisations as a powerful management tool in identifying and prioritising problem areas. Generally, COPQ

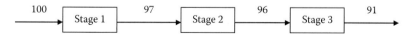

Figure 5.3 Rolled throughput yield calculation.

is estimated by organisations as a percentage of sales revenue. COPQ accounts for more than 25% of sales revenue in the context of manufacturing, and it is close to 40% in the context of many service organisations. This is estimated based on four major components, viz, internal failure cost, external failure cost, appraisal cost and prevention cost.

Internal failure cost: The cost associated with not meeting the specified requirements for products, components, materials, or services is considered as internal failure cost. These failures are discovered before the product is shipped to the customer. The major categories of internal failure cost are related to rework, scrap, retest, downtime, yield loss, downgrading of components and products, etc.

External failure cost: These are costs associated with failure of products or services after delivery to customers. These include charges for warranty, returned product, liability, loss of good will, damaging the brand image or company reputation, etc.

Appraisal cost: Appraisal costs are those costs associated with measuring, evaluating or inspecting products and components to ensure conformance to specified requirements. These include product inspection/test-related cost, instrument maintenance/calibration-related cost, product life testing, etc.

Prevention cost: Prevention costs are those costs associated with product/service design and manufacturing that are directed towards the prevention of non-conformance. Generally, these are the costs related to making things 'right first time'. The major categories under this are product/process design, process control, training, data collection and analysis, development of quality management system standards, etc.

5.3.4.1 Example 1

For a typical month, a manufacturing company identified and reported the following quality costs:

Inspection wages	$12,000
Quality planning	$4,000
Final product test	$110,000
Retest and troubleshooting	$39,000
Field warranty cost	$205,000
In-plant scrap and rework	$88,000

What is the total failure cost for this month?

The total failure cost is the sum of internal and external failure costs. The internal failure costs in the above example are retest and troubleshooting as well as in-plant scrap and rework, whereas field warranty cost is a typical external failure cost. The total failure cost = $39,000 + $205,000 + $88,000 = $332,000.

5.3.4.2   Example 2

A company incurred the following quality costs in the last month.

Evaluating finished product	$50,000
Quality auditing training	$10,000
Manufacturability study	$1,000
Repairing finished product	$5,000
Process capability studies	$30,000

How much did the company spend on prevention costs?

The prevention costs in the above case include quality auditing training, manufacturability study and process capability studies. Therefore, the total prevention costs for the above given example = $10,000 + $1,000 + $30,000 = $41,000.

5.3.5   Process capability indices

Any process will have variation in its output, and it is common for specification limits to be defined such that if the measured output of the process exceeds the specified limits, the process is deemed to have produced a non-conforming item. Process capability is a measure of how well a given process meets the customer specifications. PCIs are used to evaluate how well the process is meeting its customer specifications. The most commonly used process capability indices are C_p and C_{pk}. One of the stringent assumptions for the calculation of C_p and C_{pk} is that the process should be in a state of statistical control.

The process capability ratio, C_p, indicates how well the process distribution fits within its specification limits, and is simply the ratio of the specification width to the width of the process. C_p refers to the potential capability of a process. For any process with upper specification limit (USL), lower specification limit (LSL) and standard deviation σ, the C_p is defined as

$$C_p = \frac{USL - LSL}{6\sigma} \tag{5.1}$$

The problem with C_p is that it does not take account of how well the process distribution is centred within its limits, which can result in a process with high C_p and can have many rejects. The solution to this problem is addressed in C_{pk}, which measures a similar ratio by the following formula, where $\bar{X}$ is the average of the collected data. C_{pk} measures the actual capability of the process, and it takes account of both process centring and process variability.

$$C_{pk} = \text{Min}\left(\frac{\text{USL} - \bar{X}}{3\sigma}, \frac{\bar{X} - \text{LSL}}{3\sigma}\right) \tag{5.2}$$

A high value for C_p and C_{pk} indicates that the process variability is low, and there is no process shift. We recommend C_p and C_{pk} values to be greater than 1. However, in real-life scenarios, these values are not identical due to a shift in the process mean over the long term. A Six Sigma process has a C_p and C_{pk} value of 2 with no shift in the process mean. However, with a shift of 1.5 sigma, $C_p = 2$ and $C_{pk} = 1.5$. If C_p and C_{pk} values are identical, we can then conclude that the process is centred.

5.3.5.1   Example 1

The diameter in inches of a sample of 40 ball bearings manufactured by a certain process has an average of 0.7346 with a standard deviation of 0.0049. If the specification for this characteristic is 0.734 ± 0.02, calculate the C_p and C_{pk} values.

$$C_p = \frac{0.754 - 0.714}{6 \times 0.0049} = 1.36$$

$$C_{pk} = \text{Min}\left\{\frac{0.754 - 0.7346}{3 \times 0.0049}, \frac{0.7346 - 0.714}{3 \times 0.0049}\right\}$$

$$= \text{Min}(1.32, 1.4) = 1.32$$

As both C_p and C_{pk} values are closer to each other, we may conclude that the process is more or less centred. Moreover, the process is capable of meeting the specifications.

5.3.5.2   Example 2

A pharmaceutical company producing vitamin capsules desires a proportion of calcium content between 40 ppm and 55 ppm. A random sample of 20 capsules chosen from the output yields a sample mean calcium content of 44 ppm with a standard deviation of 3 ppm. Comment on the ability of the process to meet specifications.

$$C_p = \frac{55 - 40}{6 \times 3} = 0.833$$

$$C_{pk} = \text{Min}\left\{\frac{55 - 44}{3 \times 3}, \frac{44 - 40}{3 \times 3}\right\} = (1.22, 0.44) = 0.44$$

In this case, the process is not centred as C_p and C_{pk} values are significantly different. Moreover, both values are less than 1, and therefore it is safe to conclude that the process is not able to meet the required specifications. One should look into the possibility of reducing process standard deviation further and try to determine the process parameters which influence the process mean so that it can be fine-tuned to bring it on target (47.5 ppm, middle value of the given specifications).

5.3.5.2.1 Process performance indices The Automotive Industry Action Group (AIAG) suggests usage of process capability indices C_p and C_{pk} only when the process is stable. It recommends the usage of process performance indices, P_p and P_{pk} when the process is unstable. The P_p and P_{pk} are defined as follows:

$$P_p = \frac{USL - LSL}{6s} \tag{5.3}$$

$$P_{pk} = Min\left(\frac{USL - \bar{X}}{3s}, \frac{\bar{X} - LSL}{3s}\right) \tag{5.4}$$

where 's' is the sample standard deviation and can be determined using the following equation:

$$s = \sqrt{\frac{\sum_{i=1}^{n}(x_i - \bar{x})^2}{n-1}} \tag{5.5}$$

5.4 Lean Six Sigma metrics

Appropriate metrics are used in LSS projects to measure the process and project outcomes, detect improvement opportunities and monitor changes. These metrics will help to pinpoint sources of waste and variability and determine the root causes of problems. Business improvement strategies focus on key improvement initiatives, and deploy the right resources, tools and methodologies to utilise opportunities. Selection of the right performance metrics enables this aspect. The selected metrics depend on the project goals. The metrics can be grouped under four categories: time, cost, quality and output metrics. Time-based metrics include lead time, cycle time, takt time, response time, activity ratio, percent on-time delivery, value-added time and non-value-added time. Cost metrics include total process cost, cost savings, labour savings, etc. Quality metrics include customer satisfaction, defect rate, first pass yield, process capability indices, COPQ, etc. Output metrics include production, WIP and inventory. Other metrics include bottom-line impact, profit margin, ROI, market growth, revenue growth and so on. These measures need to be assessed after implementation of pilot LSS projects. The metrics may also be categorised in terms of cost, quality and delivery, which serve as the fundamental drivers for an organisation. Also, the categorisation could be based on process and organisation dimensions. The metrics serve as the base for quantification and utilisation of improvement opportunities. Also, the metrics evaluate potential process improvements and prioritise appropriate actions, establish baseline data for process improvement and communicate the results of LSS implementation.

5.5   Overall equipment effectiveness

5.5.1   Why do we need OEE?

Overall equipment effectiveness (OEE) helps organisations to

- Determine the best avenues to monitor and improve the effectiveness of manufacturing processes.
- Consider various sub-parameters of the manufacturing process: availability, performance and quality. The three factors are multiplied, and product is expressed as a percentage.
- Evaluate it as a key performance metric in Lean manufacturing and total productive maintenance (TPM) practices. It is a consistent way to ensure the effectiveness of TPM and other initiatives by contributing an overall framework for evaluating production efficiency.

The definition of OEE does not consider all factors that reduce capacity utilisation, for example, planned downtime, lack of material input, lack of labour, etc. (Nakaiima 1988). TPM depicts the relationship between production and maintenance, for ensuring continuous improvement in terms of product quality, operational efficiency, capacity, assurance and safety (Nakaiima 1988).

Nakaiima (1988) emphasises that OEE computation is an effective approach for analysing the efficiency of a manufacturing system, and it is expressed as a function of availability, performance efficiency and quality. Availability losses happen because of breakdowns. Performance losses occur because of line stoppages. Quality losses occur due to product rejects.

There are three general loss categories to be considered while computing OEE. They are downtime loss, speed loss and quality loss. OEE for world-class organisations must be 85% or greater, and the OEE rate of manufacturing plants is 60% (Ahuja and Khamba 2008).

5.5.2   Availability

It is essential to compare the potential operating time and the actual time in which manufacturing products or providing services is essential. Availability takes downtime loss into account, which includes any events or activities that prevent planned production for an appreciable time. Examples include breakdowns, minor stoppage, equipment failures, material shortages and changeover time. Changeover time is considered in OEE analysis as it is a representative of downtime. Changeover time cannot be eliminated but can be reduced in most cases using tools such as SMED (readers may refer to Chapter 7 for more information on SMED). The remaining available time is called the operating time.

$$\text{Availability} = \frac{\text{Operating time}}{\text{Planned production time}} \times 100\% \qquad (5.6)$$

5.5.3   Performance (utilization or speed)

Performance is the comparison of the speed or actual output with which the system could be consistently manufacturing in the same time frame. Performance takes speed loss into account, which includes factors that make the process operate at less than the maximum possible speed. Examples include idling, machine wear and operator inefficiency. The remaining available time is considered as the net operating time.

$$\text{Utilization} = \frac{\text{Actual time}}{\text{Maximum output in running time}} \times 100\% \qquad (5.7)$$

$$\text{Speed} = \frac{\text{Actual speed}}{\text{Best demonstrated speed}} \times 100\% \qquad (5.8)$$

5.5.4   Yield (quality)

It is required to compare the raw materials quantity fed to the process and the number of products/services that meets customer's specifications. Quality considers loss, which considers the produced quantity that does not comply with quality standards, including products that need rework. Table 5.4 presents OEE loss, drivers and computation measures.

$$\text{Yield} = \frac{\text{Actual output} - \text{defects}}{\text{Actual input}} \times 100\% \qquad (5.9)$$

Table 5.4 OEE loss, drivers and computation measure

OEE loss	OEE drivers	Calculation	Achievable
Downtime loss	Availability	Ratio of operating time to planned production time	Process must run without any stoppage
Speed loss	Performance	Ratio of actual output to maximum output in running time	Process must run at its theoretical maximum speed
Quality loss	Quality	Ratio of good pieces to total pieces	There must be no rejected or reworked products

5.5.5   Six big losses

The loss of efficiency in manufacturing processes is mainly due to six big losses. The scope of OEE and TPM is to reduce or eliminate six big losses which are responsible for reducing equipment effectiveness. The six big losses are (Dal et al. 2000)

1. *Breakdown*: This loss is mainly due to equipment defects, which result in reduced speed of the manufacturing process. It is categorized as downtime loss as it reduces output. It is mainly caused by equipment failure, tool failure and unplanned maintenance. It can be avoided by conducting frequent maintenance programs on operating equipment and tools as per the specified working conditions.
2. *Set-up and adjustments*: This loss is due to changes in operating conditions such as changeovers, set-up of tools and performing trial-and-error adjustments. It results in time loss as more time is spent on changeovers and initial adjustments and is categorised as downtime loss. It can be remedied by conducting time reduction programs.
3. *Minor stoppages*: It is mainly due to idling of machines. The reasons for small stops include misfeeds, component jams and interrupted product flow. It is categorised as speed loss since there is a block in the production flow. Minor stoppages only include minor stops that are less than 9 minutes and do not need help from maintenance personnel to solve them.
4. *Reduced speed*: This loss includes losses that occur due to slow equipment speed and losses due to difference in current machine speed and actual machine speed. It is recommended to run the equipment at its theoretical running speed. At higher operating speeds, quality defects and minor stoppages happen frequently, and thereby it is required to operate the machine at a lower, more moderate speed. Speed losses are measured in terms of the ratio of theoretical to actual operating speed.

5. *Start-up rejects*: This loss is due to the manufacture of defective products during warm-up and early stages of operation. These products must be reworked or scrapped. It is categorised as quality loss and is measured as the ratio of the number of quality products to total production.
6. *Production rejects*: These are the rejects that are produced during the steady-state production. This loss is categorised as a quality loss as it requires rework.

5.5.6   Calculating OEE

OEE is calculated as the product of availability, performance and quality (Kwon and Lee 2004).

5.5.6.1   Availability
Availability is categorised as downtime loss and is calculated as

$$\text{Availability} = \frac{\text{Operating time}}{\text{Planned production time}} \tag{5.10}$$

5.5.6.2   Performance
Performance is categorised as speed loss and is calculated as

$$\text{Performance} = \frac{\left\{\dfrac{\text{Total units}}{\text{Operating time}}\right\}}{\text{Ideal run rate}} \tag{5.11}$$

5.5.6.3   Quality
Quality is categorised as quality loss and is calculated as

$$\text{Quality} = \frac{\text{Good pieces}}{\text{Total pieces}} \tag{5.12}$$

5.5.6.4   OEE
OEE is calculated by considering all three factors and is computed as the product of availability, performance and quality.

$$\text{OEE} = \text{Availability} \times \text{Performance} \times \text{Quality} \tag{5.13}$$

5.5.6.5   Illustrative example
Table 5.5 contains the data needed for OEE computation including the calculation of OEE factors: availability, performance and quality. The same

Table 5.5 Data for OEE computation

Parameter	Data
Shift length	8 hours = 480 minutes
Short breaks	2 @ 10 minutes = 20 minutes
Lunch break	1 @ 30 minutes = 30 minutes
Downtime	45 minutes
Ideal run rate	70 pieces/minute
Total pieces	20,000 pieces
Reject pieces	400 pieces

units of measurements as mentioned (minutes and units) must be followed throughout the calculation to ensure consistency.

Planned production time = [Shift length − Breaks] = [480 − 50] = 430 minutes

Operating time = [Planned production time − Downtime] = [430 − 45] = 385 minutes

Good pieces = [Total pieces − Reject pieces] = [20,000 − 400] = 19,600 pieces

$$\text{Availability} = \frac{385}{430}$$

$$= 0.8953$$

$$= 89.53\%$$

$$\text{Performance} = \frac{\left\{ \dfrac{20,000}{385} \right\}}{70}$$

$$= 0.7421$$

$$= 74.21\%$$

$$\text{Quality} = \frac{19,600}{20,000}$$

$$= 0.98$$

$$= 98\%$$

$$\text{OEE} = \text{Availability} \times \text{Performance} \times \text{Quality}$$

$$= 0.8953 \times 0.7421 \times 0.98$$

$$= 0.6511$$

$$\text{OEE} \cong 65\%$$

Although the quality part is good, the availability and performance aspects can be further improved to increase the OEE values further.

5.6   Summary

This chapter highlights the need for and benefits of identifying LSS performance metrics for SMEs. The guidelines for selecting the LSS metrics include the identification of the purpose for the usage of metrics, usage of few appropriate metrics rather than too many metrics and focus on all stakeholders. The common metrics of Lean include takt time, cycle time, lead time, changeover time, etc. The common metrics of Six Sigma include DPMO, sigma quality level, throughput yield, rolled throughput yield, COPQ and PCIs. The common metrics of LSS, especially the financial metrics, are also highlighted in the chapter. Also, the importance of OEE, along with an illustrative example, are presented for a good understanding for the readers.

References

Ahuja, I. P. S. and Khamba, J. S. (2008). An evaluation of TPM initiatives in Indian industry for enhanced manufacturing performance. *International Journal of Quality and Reliability Management* 25(2): 147–172.

Dal, B., Tugwell, P. and Greatbanks, R. (2000). Overall equipment effectiveness as a measure of operational improvement: A practical analysis. *International Journal of Operations and Production Management* 20(12): 1488–1502.

Devadasan, S. R., Sivakumar, V. M., Murugesh, R. and Shalij, P. R. (2012). *Lean and Agile Manufacturing: Theoretical, Practical and Research Futurities.* New Delhi: PHI Learning.

Gopalakrishnan, N. (2010). *Simplified Lean Manufacture: Elements, Rules, Tools and Implementation.* New Delhi: PHI Learning.

Kavanagh, S. and Krings, D. (2011). The 8 sources of waste and how to eliminate them. *Government Finance Review* 27: 6–18.

Kwon, O. and Lee, H. (2004). Calculation methodology for contributive managerial effect by OEE as a result of TPM activities. *Journal of Quality in Maintenance Engineering* 10(4): 263–272.

Nakaiima, S. (1988). *Introduction to Total Productive Maintenance (TPM).* Cambridge, MA: Productivity Press (translated into English from the original text published by the Japan Institute for Plant Maintenance, Tokyo, 1984).

Tapping, D., Luyster, T. and Shuker, T. (2002). *Value Stream Management: Eight Steps to Planning, Mapping, and Sustaining Lean Improvements.* New York: Productivity Press.

chapter six

Six Sigma methodology

6.1 Introduction

Six Sigma utilises a five-stage methodology (define-measure-analyse-improve-control – DMAIC) for tackling problems related to existing processes. The purpose of using this powerful problem-solving methodology is to understand and evaluate the root causes of a given problem. DMAIC is an iterative process that gives structure and guidance to improving processes in any workplace. The five steps in the Six Sigma methodology are easy to understand, and they are logical in their sequence. Moreover the steps allow a team to adequately scope the problem, measure the current performance by quantifying the problem, analyse the root causes of problems, test and verify improvement recommendations and then implement changes for sustainability over the long haul (Brassard et al. 2002).

DMAIC methodology is extremely useful in two scenarios:

1. *Complex problems*: In complex problems, the causes and solutions are not obvious. To determine the root causes of a problem, you need to bring together people with different levels of skills, knowledge and experience.
2. *Solution risks are high*: We recommend organisations to utilise DMAIC any time when the risks of implementation are high, even if the solution is obvious and straightforward.

It is always risky to skip any DMAIC steps. The success of DMAIC lies with the logical sequence of steps in the methodology. However, it is human nature to want to jump to solutions and quickly make the improvement. If you think you have a solution to the problem at hand, and if the solution is obvious, you can then try to skip some of the DMAIC steps. However, it is important to ask the following questions before you make a decision on whether you need to skip DMAIC steps or not:

- Do I have data to show that the proposed solution to the problem is the best? In other words, what evidence do I have to justify that the solution to the problem at hand is the best?
- How do I know that the solution will really solve the given problem at hand?

6.2   Define phase

The goal of the *define* step is to define the scope of the project and obtain background information about the process where the problem lies and its customers (internal or external or both). We recommend the following points to be taken into account in the *define* phase of the project:

- *Develop and review project charter*: It is important to develop a project charter in the define phase and make sure that the project champion reviews it and provides constructive feedback to the charter. Chapter 7 (refer to Section 7.14) provides further explanation of this powerful tool along with a template. A project charter includes the voice of the customer (VOC), critical-to-quality (CTQ) characteristics and the specifications for CTQs. The readers are encouraged to refer to Sections 7.13 and 7.14 for more information on VOC and CTQs.
- *Problem statement and goals*: Review existing data to confirm that the problem exists, and ensure that it is important to your customers (internal, external or both) as well as business.
- *Validate financial benefits*: It is critical to understand the potential benefits (both financial and non-financial) and the total investment to the project. We recommend a minimum of 1 to 3 for the investment-to-benefit ratio.
- *High-level process map*: We recommend the development of a high-level process map (also known as SIPOC) to verify the project scope and to understand the key inputs, outputs and the processes at a high level. SIPOC stands for suppliers, inputs, processes, outputs and customers. Please refer to Chapter 7 (Section 7.1) for more information on SIPOC.
- *Develop project plans*: A project plan should include the five phases: key milestones, schedule, team members, key deliverables and potential risks associated with the project.
- *Communication plan*: Regular communication with project team members and stakeholders (sponsors, customers, managers, etc.) can help everyone to understand the progress of the project and create more buy-in and avoid any pitfalls in the execution of the project.
- *Tollgate review*: A formal review process that helps keep the project on track. The Lean Six Sigma (LSS) champion should be executing the tollgate review, and he or she needs to approve the key findings of the *define* phase before the project leader and his or her team members can move on to the measure phase.

The following tools may be used in the *define* phase of the methodology. It is important to note that these tools are explained in Chapter 7. We recommend readers to refer to Chapter 7 for more information about the use of LSS tools.

Recommended tools in the *define* phase include

- SIPOC
- Value stream mapping (VSM)
- Project charter
- VOC analysis

6.3   *Measure phase*

The goal of the *measure* step is to quantify the problem by gathering information about the current situation. The key outputs of the *measure* phase include

- Data that pinpoint the problem's location in the process
- Baseline data showing how well the process meets customer needs
- A more focused problem statement

We recommend the following points to be taken into account in the *measure* phase of the project:

- *Collect baseline data*: The baseline performance of the process must be established in the measure phase. The typical measures include defect rate (defects per million opportunities – DPMO), process capability indices (C_p/C_{pk}), sigma quality level (SQL) of the process, throughput yield or rolled throughput yield, process lead time, cycle time of process, etc. (refer to Chapter 5 for more information). In some cases, it can be the cost of poor quality or even different forms of waste based on waste analysis.
- *Create a data collection plan*: In the data collection planning, one should think about the type of data to be collected, how the data are relevant to the problem at hand, how many samples should be collected from the process and how frequently these samples should be collected.
- *Check the accuracy and precision of data*: We strongly recommend one to execute a gage repeatability and reproducibility study (measurement system analysis or MSA) to ensure that the measurement system is valid and sound. This will ensure the quality and reliability of data collected from the process. For more information on MSA, readers may refer to Chapter 7 (see Section 7.16).
- *Operational definition for defects*: In the measure phase, we need to understand what to measure (also called CTQs), and once we identify the CTQs, we need to understand the operational definition for

defects. What constitutes a defect should be understood by all team members, and there should be a standard definition for the term 'defect'.

- *Use of run charts and control charts*: It is a good practice to construct a run chart or control chart based on the baseline data. For more information on these charts, please refer to Chapter 7 (Sections 7.11 and 7.12).
- *Tollgate review*: We recommend the team leader and members to pursue a tollgate review before moving on to the *analyse* phase.

Recommended tools in the *measure* phase include

- CTQs and CTQ tree
- Detailed process mapping
- Run charts
- Control charts
- MSA
- Process capability analysis (refer to Chapter 5)

6.4   Analyse phase

The goal of the *analyse* step is to identify root causes and confirm them with data. In many cases, we determine the key input variables of the process, perform data analysis, determine root causes and prioritise root causes.

The key outputs of the *analyse* phase include

- Identification of potential causes associated with the problem
- Understanding of potential input variables (Xs) which influence the CTQs or (Ys)
- Identification of value-added and non-valued-added steps/ activities

We recommend the following points to be taken into account in the *analyse* phase of the project:

- *Conduct 7 +1 forms of waste analysis*: Analyse the different forms of waste in our process under consideration.
- *Analyse the process flow*: Understand the bottlenecks in the process, assess the value-added and non-value-added activities in the process from a customer's perspective, analyse the CTQs, understand the key drivers of CTQs, etc.
- *Analyse data collected in measure phase*: Evaluate the relationship between the effect of the problem and the potential causes which

lead to the problem, use relevant tools (Pareto analysis, hypothesis testing, root cause analysis, etc.) to narrow the search for root causes.

- *Verify significant relationships between causes*: Use scatter plots or more sophisticated tools (e.g. correlation analysis, regression analysis, etc.) to verify significant relationships between causes.
- *Tollgate review*: A tollgate review must be executed in the *analyse* phase before one can move on to the *improve* phase.

Recommended tools in the *analyse* phase include (George et al. 2005)

- Failure mode and effect analysis
- Scatter plot
- Correlation analysis
- Cause and effect analysis or fishbone analysis
- Root cause analysis
- Pareto analysis
- Regression analysis
- Histogram
- Hypothesis tests

6.5 Improve phase

The goal of the *improve* step is to develop, try out and implement solutions that address root causes. In this phase, one may generate potential solutions, select and prioritise solutions, perform risk assessment, pilot the solution for its effectiveness and finally evaluate the benefits.

The key outputs of the *improve* phase include

- Planned and effective actions that can eliminate or reduce the impact of the problem at hand
- 'Before' and 'after' analysis that shows how much of the performance improvement gap was closed

The improve phase not only involves deriving solutions for the problem at hand but also using the plan-do-check-act (PDCA) cycle to evaluate and improve the solutions you want to implement. We recommend the following points to be taken into account in the *improve* phase of the project:

- *Develop potential solutions*: Potential solutions will be derived in the *improve* phase. This is one step where pushing for creativity is highly desired.

- *Evaluate and optimise the possible best solution*: It is recommended to develop criteria before choosing an appropriate solution. The criteria may include ease of implementation, costs associated with implementation, associated risks, etc.
- *Implement pilot solution*: The best solution derived for the problem needs to be piloted.
- *Compare before and after scenarios*: In this step, one has to compare the baseline performance data with improved performance data to make sure that the solution is effective.
- *Tollgate review*: A tollgate review must be executed in the *improve* phase before one can move on to the *control* phase.

Recommended tools in the *improve* phase include

- Brainstorming
- Failure modes and effect analysis
- Solution prioritisation matrix
- Design of experiments
- Kanban system
- Single minute exchange of dies

6.6   Control phase

The goal of the *control* step is to sustain the gains by standardising the work methods or processes, anticipating future improvements and capturing and documenting the key lessons learned from the project and exploring the opportunities of transferring the knowledge to other operations in the business. In this phase, one should document standard operating procedures, develop process control plans, create project storyboard and transit process ownership.

The key outputs of the *control* phase include

- Documentation of the new process or method
- Training of employees in the new method or process
- A system for monitoring the implemented solution (process control plan) along with specific metrics to be used for regular process auditing
- Completed project documentation, including lessons learned and recommendations for further opportunities

We recommend the following points to be taken into account in the *control* phase of the project:

- *Documenting and standardising the improvements*: The first step of the control phase is to document and standardise the improvements that were rolled out during the *improve* phase.
- *Creating a process monitoring plan*: Perhaps the most critical aspect of the control phase is establishing a plan to monitor the new process and act when results are not up to specifications so that the project gains will be maintained.
- *Control charts to monitor the improved process performance*: We strongly recommend the use of control charts for evaluating process stability over a period of time.
- *Six Sigma storyboard*: A storyboard is a display created and maintained by a project or process improvement team that tells the story of a project. It is a powerful communication mechanism to the team and stakeholders. It may be used in the future as a guide for other or similar projects.
- *Use of PDCA cycle*: PDCA cycle serves as a reminder to think of improvement as being continual: How do we improve the process further from where we are now? What areas of the process need to be improved further and what types of projects need to be pursued in the future?

Recommended tools in the *control* phase include

- Run chart
- Control chart
- 5S practice
- Visual management
- Kanban system
- PDCA cycle
- Poka-Yoke (mistake proofing)

6.7   Summary

This chapter provides readers with an introduction to Six Sigma problem-solving methodology (DMAIC). The authors would like to accentuate the point that this methodology should be used only for fixing problems in the existing processes. It is strongly recommended not to employ this methodology for tackling poor design problems in your existing processes. The authors have explained the five phases of the Six Sigma methodology in a cookbook fashion and provided the relevant tools within each phase which guide the process improvement leaders or project leaders through problem-solving scenarios. The detailed explanation of these tools is provided in Chapter 7.

References

Brassard, M., Finn, L., Ginn, D. and Ritter, D. (2002). *The Six Sigma Memory Jogger II, A Pocket Guide of Tools for Six Sigma Improvement Teams*. New York: GOAL/QPC.

George, M. L., Maxey, J., Rowlands, D. T. and Price, M. (2005). *The Lean Six Sigma Pocket Toolbook: A Quick Reference Guide to Nearly 100 Tools for Improving Quality, Speed, and Complexity*. New York: McGraw-Hill.

chapter seven

Basic and advanced tools of Lean and Six Sigma for SMEs

7.1 Suppliers, Inputs, Process, Outputs and Customer

7.1.1 What is SIPOC?

SIPOC is a process improvement tool that provides a key summary of the inputs and outputs of one or more processes in tabular form. The acronym SIPOC denotes suppliers, inputs, process, outputs and customers, which represent the columns of the table. It was applied in the field of total quality management in the 1980s and widely used in Six Sigma, Lean manufacturing, and other business process improvement strategies. SIPOC is a vital tool for documenting a business process from start to end. SIPOC is used in the define phase of the define-measure-analyse-improve-control (DMAIC) process (Adams et al. 2004).

The uses of SIPOC include

- Provide a detailed overview of processes
- Enable the definition of a new process

7.1.2 How do you construct a SIPOC diagram?

A SIPOC diagram represents the high-level view of a process. The SIPOC diagram plays an important role in process definition/improvement. It may be used by the analyst in collaboration with other stakeholders to arrive at a consensus on the process before moving to a higher level of detail (Evans and Lindsay 2005).

7.1.3 When do you use a SIPOC diagram?

- To define the scope of the project.
- To document/assess an existing process prior to the deployment of improvement efforts.
- To obtain a high-level understanding of a process during the initiation of a process improvement activity. It helps the process owner

and those related to the process to arrive at a consensus on process boundaries.
- To discuss the process and obtain agreement before drawing process maps.

7.1.4   How do we create a SIPOC diagram?

- *Step 1*: Designate the process.
- *Step 2*: Indicate the start/end as well the process scope.
- *Step 3*: Depict the output(s) of the process.
- *Step 4*: Indicate the customer(s) of the process.
- *Step 5*: Depict the supplier(s) of the process.
- *Step 6*: Define the input(s) of the process.
- *Step 7*: Indicate present highest level of process steps.

7.1.4.1   General rules for drawing a SIPOC diagram (Pyzdek and Keller 2003)

- Maintain it as high level and as simple as possible.
- It needs to be organised using a brainstorming session to generate ideas on the elements.
- *Suppliers*: Provide inputs.
- *Inputs*: Are the key requirements required for the process to work.
- *Outputs*: Are the results of process steps.
- *Customers*: Receive or use the outputs of the process.

7.1.4.2   Practical application

1. Organise a brainstorming session including participants who are knowledgeable about the process.
2. Develop a diagrammatic representation of the process on a display board considering inputs from all members.
3. *Identify the outputs*: What are the outcome(s) of each step?
4. From the outputs, identify the customers.
5. For each step in the process, determine what inputs are worked on to execute the process.
6. *Suppliers*: Who are those that provide inputs to the process?

7.1.5   An illustrative example

A pump manufacturing firm is interested to increase its productivity and decrease its defect ratio by adopting the DMAIC approach. As a part of the define phase, SIPOC was developed to study all elements and members involved in pump manufacturing. It helped in defining the scope and boundaries of the project. The SIPOC diagram developed is shown in Figure 7.1.

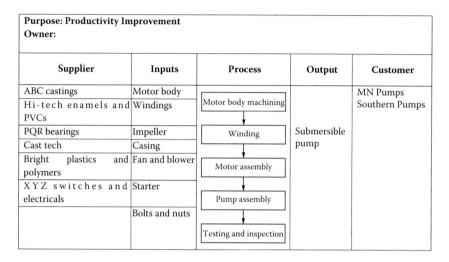

Figure 7.1 SIPOC diagram.

7.2   Value Stream Mapping

7.2.1   What is VSM?

- VSM is a Lean technique used for analysis of the current state and deriving a desired future state for the series of processes that take a product or service from its start through to the customer (Tapping et al. 2002).
- VSM is a Lean manufacturing technique used to document, analyse and improve the information flow or materials flow required to develop a product or service for a customer.
- VSM is a paper-and-pencil tool that helps to observe and analyse the material and information flow as a product or service developed through the value stream.
- VSM may be used in the measure, analyse and improve phases of the DMAIC process.

7.2.2   Why do you use a VSM tool? (Hines and Rich 1997)

- Graphically depict, analyse and perceive the flow of materials and information needed to process them.
- Depict the interaction between multiple functions within the manufacturing process and their associated functions.
- Represent the flow of information (communications) and materials throughout the complete manufacturing process.
- Signal problems, inefficiencies and losses within complex systems.

- Develop and implement improvements in a highly visual manner that facilitates culture change within the organisation.
- Provide guidelines for Lean transformation teams and top management towards continuous improvement.
- Enable as a dashboard to track and continuously improve the process.

7.2.3   *When do you use value stream maps?*

VSM is to be used for a high-production, low-variety product mix with few components.

A current state map depicts the current scenario of processes. This is required both to understand the need for change and to plan the improvement.

7.2.4   *How do we create a VSM?*

Train the VSM team: Form a cross-functional team that encompasses all stakeholders of the process to be mapped. It must include operators and maintenance personnel who possess knowledge on the processes.

Undertake a physical walk through the shop floor to map out the production process.

Document each step and identify the communication channels.

Create the 'current state' VSM with all needed data and information. Revise the VSM until the description of the current process is accurate and complete.

7.2.4.1   *Step-by-step procedure*

Step 1: Select the team leader
Step 2: Form the cross-functional Lean team
Step 3: Select the process to be mapped
Step 4: Collect data and generate current state map
Step 5: Analyse current state
Step 6: Map desired future state
Step 7: Develop action plans and focus on deployment
Step 8: Quantify benefits

7.2.5   *An illustrative example*

A manufacturing firm is interested in implementing the Lean technique in its product line, which manufactures piston valves. Cutting is the first operation to be performed on the raw material followed by turning and drilling operations. Next, the machined component is allowed to harden, and then grinding is performed. Finally the manufactured camshaft is inspected and sent for packaging and dispatching. A total of 11 workers are involved in manufacturing the camshafts, and their tasks are properly assigned. The firm operates 8 hours per day (including a 30-minute lunch break and a 10-minute tea break) and manufactures 150 piston valves per day. The individual cycle time for each process was identified by conducting a time study, and work-in-process (WIP) inventory data were also collected visually. Based on the available data, total cycle time and lead time was calculated and found to be 56 minutes and 8.71 days. Takt time was found to be 4.3 minutes for the current state. The developed current and future state maps are shown in Figures 7.2 and 7.3.

After analysing the current state, necessary improvement actions were planned. Improvements were made with reference to the individual processes, and improvements were observed. After successful implementation of the improvement actions, a future state map was constructed.

7.3   5S practice

7.3.1   What is 5S?

5S is a workplace organisation method that implies five Japanese words: seiri, seiton, seiso, seiketsu and shitsuke. The translated terms are 'sort', 'straighten', 'shine', 'standardise' and 'sustain' (Rojasra and Qureshi 2013). 5S implies workplace organisation for efficiency and effectiveness by identifying and storing the items used, maintaining the area and items and sustaining orderliness. 5S is used in the improve phase of the DMAIC process.

It is one of the most powerful Lean manufacturing tools. 5S is a simple tool for ensuring an organised workplace that is clean, efficient and safe to enhance productivity and visual management and to ensure work standardisation. 5S is a team-oriented task.

Seiri: It includes removal of unnecessary items and the disposal of them properly. It evaluates necessary items based on cost and the removal of all parts not in use.

Seiton: It includes the arrangement of all necessary items for easy retrieval to prevent time loss and enable smooth workflow.

Seiso: Ensure cleanliness and safety of workplace.

Seiketsu: Standardise best practices in the work area by ensuring high housekeeping standards and orderliness.

Shitsuke: Ensure working order by means of regular audits, training and the enforcement of the discipline.

7.3.2   Why do you use 5S?

- To eliminate waste
- To instil quality culture at the workplace
- To ensure ease of application

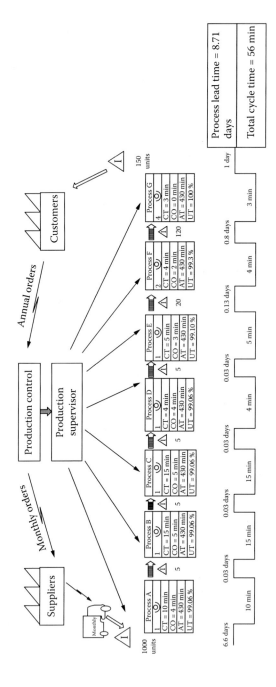

Figure 7.2 Current state map.

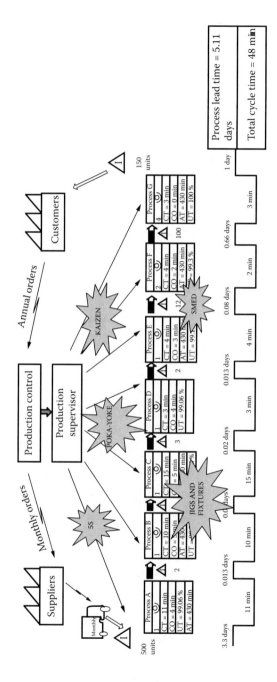

Figure 7.3 Future state map.

The main benefits of implementing 5S include

- To ensure a cleaner, safer, well-organised workplace
- To improve floor space utilisation
- To ensure a smoother and a more systematic workplace
- To reduce machine breakdowns
- To make defect-free products
- To improve morale and satisfaction of employees
- To improve productivity of the organisation

7.3.3   What is involved in 5S?

Sorting: Establish criteria and identify items not in use, and attach red tags to such items. The red tagged items are categorised and moved to a designated space. A team is formed to identify disposal actions and those actions are recorded.

Simplification: Determine the location for each item. A shadow board should be developed, and items labelled. The layout and equipment are then determined.

Systematic cleaning: Systematic cleaning enables a way to inspect, by performing a clean sweep around a work area. Check points should be identified for examining performance, and acceptable performance can then be determined. Visual indicators are used to mark equipment and control checks to be done.

Standardising: Standardisation implies that everyone knows what is expected. A routine check sheet should be developed for each work area. An audit system must be developed to document standard methods.

Sustaining: It is necessary to sustain the improvements. 5S level of achievement is to be ascertained. Worker-led routine should be done using the 5S checklist. The scheduled routine checks to be done followed by higher-level audits.

7.3.4   An illustrative example

Table 7.1 shows the audit sheet developed for evaluating 5S practice.

7.4   Single Minute Exchange of Dies

7.4.1   What is SMED? (Moreira and Pais 2011)

SMED stands for single minute exchange of dies. SMED includes the deployment of a theory and set of techniques that enable equipment

Table 7.1 5S audit sheet

5S Audit Sheet

Name:
Date:
Team:

For each question, mark the score that best represents the occurrence of the observations taken.

1=No evidence shown, **2=Some** evidence in areas, **3=Good** evidence in areas, **4=Extensive** evidence in areas, **5=Found everywhere**, no exceptions (100%)

Sort	Very poor	Poor	Good	Very good	Excellent
1. Is the area free from unnecessary inventory/WIP?	1 Details:	2	3	4	5
2. Is the area free from any unnecessary/stationary materials?	1 Details:	2	3	4	5
3. Is the area free from documents that are not essential, duplicate and/or out-of-date documents?	1 Details:	2	3	4	5

Set in order	Very poor	Poor	Good	Very good	Excellent
1. Necessary inventory/WIP identified, location defined & stored in correct place.	1 Details:	2	3	4	5
2. Is the work area fitted with appropriate lightings?	1 Details:	2	3	4	5
3. Is there provision to clearly identify poor quality work?	1 Details:	2	3	4	5

Shine	Very poor	Poor	Good	Very good	Excellent
1. Is equipment clean & free of dust?	1 Details:	2	3	4	5
2. Are the containers free from dirt?	1 Details:	2	3	4	5
3. Are floors in the area clean, free of unnecessary items?	1 Details:	2	3	4	5

(Continued)

Table 7.1 (Continued) 5S audit sheet

Standardise	Very poor	Poor	Good	Very good	Excellent
1. Are the employees clear on the 5S implementation?	1	2	3	4	5
	Details:				
2. Can all employees detail overall responsibilities of others to 5S activity?	1	2	3	4	5
	Details:				
3. Is there evidence of understanding the importance of 5S principles?	1	2	3	4	5
	Details:				
Sustain	Very poor	Poor	Good	Very good	Excellent
1. Are 5S plans and action updates clearly depicted?	1	2	3	4	5
	Details:				
2. Are success stories displayed & confirmed for improvement?	1	2	3	4	5
	Details:				
3. Are display boards, activity charts, notice boards etc. up to date and periodically checked?	1	2	3	4	5
	Details:				
TOTAL SCORE					

set-up and changeover operations in less than 10 minutes, in the single minute range. It is not possible to attain single minute range for all set-ups, but SMED focuses on reducing set-up times. SMED is used in the improve phase of the DMAIC process.

7.4.2   *When do you use SMED?*

Modern customers demand a variety of products in arbitrary quantities needed. They expect high quality, good price and speedy delivery. SMED helps to fulfil these customer needs with reduced set-up time coupled with cost effectiveness to generate products in smaller lots.

It avoids problems associated with lot production such as inventory waste, delay and reducing quality. Benefits of SMED include quick set-ups, flexibility, faster delivery, better quality and high productivity.

7.4.3   *How do we create a SMED?*

7.4.3.1   *Stage 1: Separating internal and external set-up*
The most vital step is to separate internal and external set-ups. The internal set-up time could be reduced by performing preparatory and other

transportation tasks while the machine is being engaged. The separation of internal and external set-up tasks could be facilitated by the usage of checklists, functional checks and improving transport of die and other parts.

7.4.3.2   Stage 2: Converting internal set-up to external set-up

Further set-up times could be reduced towards single minute range by re-examining operations to determine whether any steps are wrongly designated as internal setup, and finding avenues to convert such steps to external set-up on the basis of true function.

7.4.3.3   Stage 3: Streamlining all aspects of set-up operation

To further reduce set-up time, the fundamental elements of each set-up are analysed in detail. Specific principles are applied to reduce the time needed, exclusively for steps that are to be done as internal set-up, when the machine is stopped. External set-up improvements include streamlining the storage and transport of parts and tools as well as tool and die management. When set-up is done using parallel operations, it is important to ensure reliable and safe operations and reduce waiting time. To help streamline parallel operations, workers develop and comply with procedural charts for each set-up.

7.4.4   An illustrative example

Analysis of operations is shown in Table 7.2. Categorisation of activities is shown in Table 7.3. Improvement ideas are presented in Table 7.4.

Table 7.2 Description of activities

Operation number	Description	Time required (minutes)
01	Mount the component on machine bed	10
02	Load the component	15
03	Orientation of component	35
04	Clamp the component	30
05	Indexing	48
06	Place the covers	6
07	Machining	300
08	Open the covers	6
09	Unload the component	10
10	Remove the chips	10

Table 7.3 Categorisation of internal and external activities

Operation number	Description	Activity	Time (min)
01	Mount the component on machine bed	Internal	10
02	Load the component	Internal	15
03	Orientation of component	Internal	35
04	Clamp the component	Internal	30
05	Indexing	Internal	48
06	Place the covers	Internal	6
07	Machining	External	300
08	Open the covers	Internal	6
09	Unload the component	Internal	10
10	Remove the chips	Internal	10

Table 7.4 Improvement ideas and time savings

S. no.	Activity	Before SMED	Improvement ideas	After SMED	Time saving (min)
01	Mount the component on machine bed	10	Pallet changer	0	10
02	Load the component	15	Pallet changer	0	15
03	Orientation of component	35	Poka-Yoke	26	9
04	Clamp the component	30	Pneumatic torque	10	20
05	Indexing	48	Auto-index	20	28
06	Place the covers	6	–	6	
07	Machining	300	–	300	
08	Open the covers	6	–	6	
09	Unload the component	10	Pallet changer	0	10
10	Remove the chips	10	Pallet changer	0	10
	Total	**470**		**368**	**102**

7.5 Visual management

7.5.1 What is visual management?

Visual management is described as 'management by eyes' (Murata and Katayama 2010). It includes management of activities by visual control. Visual management is a system that indicates abnormal conditions. Visual management allows us to visualise the normal state (standard) and the deviation from it. Visual management is used in the improve phase of the DMAIC process.

7.5.2 When do you use visual management?

- To make problems obvious
- To make work easier
- To indicate standard quantity
- To trigger entities that are abnormal

7.5.3 How do we perform visual management?

Three basic steps include

- Decide what entities to be made visual. A focus group or continuous improvement team within the organisation can help identify a place to start.
- Set up a way for the system to trigger a response when the system experiences an abnormal condition.
- Decide the means to respond when there is an abnormal condition.

7.5.3.1 Other industry-specific examples

1. Floor outlines or simple markings on the floor indicate the location of items.
2. A production board in manufacturing plants that displays the production numbers hour by hour and shift by shift.
3. Painting or taping up a red line on the wall or shelf at the point to indicate the reordering of items.
4. Notice boards (manual or computerised) in departments and institutes to convey information to students.
5. Seating chart in the examination hall to indicate the seating arrangement of candidates appearing for an examination.
6. Systematic tracking of operational information.
7. Communicate activities to team members on a daily basis.

7.5.4 An illustrative example

Sample visual management boards are shown in Figures 7.4 and 7.5.

7.6 Standard Operating Procedures

7.6.1 What is SOP?

A SOP is specific to the operation and depicts the activities needed to complete tasks in line with industry regulations, and business standards

Hour by Hour board				Date: Demand: Takt time:
Time	Plan	Actual	Stoppage time	Reason
7:00 am – 8:00 am	20 nos	17 nos	NIL	
8:00 am – 9.00 am	20 nos	18 nos	NIL	
....				
17:00 pm – 18:00 pm	20 nos	15 nos	10 min	Machine break down

Figure 7.4 Visual management board to monitor hourly production.

(Doolen and Hacker 2005). SOPs play a vital role in business operations. SOPs are policies, procedures and standards that are needed in operations, marketing and administration disciplines to ensure business success. A SOP is used in the control phase of the DMAIC process.

The objectives of developing a SOP include

- To improve efficiency and profitability
- To ensure consistency and reliability in production and service
- To ensure a healthy and safe environment

7.6.2   *When do you use an SOP?*

Development and implementation of effective SOPs is needed for today's competitive business environment:

1. To ensure consistency so as to achieve top performance
2. To reduce system variation so as to improve production efficiency and quality control
3. To facilitate training
4. To help in conducting performance evaluations

7.6.3   *How do we create an SOP?*

1. Designate the SOP using descriptive action words, and ensure that the SOP must be accessed by all employees.
2. Indicate the scope of the SOP, detailing specific operations or tasks, coverage of operations and purpose of the SOP.
3. Generate an overall task description indicating the people requirement, their skill set, equipment and supplies required, safety equipment required and description of end product.

Illustrative Example 2

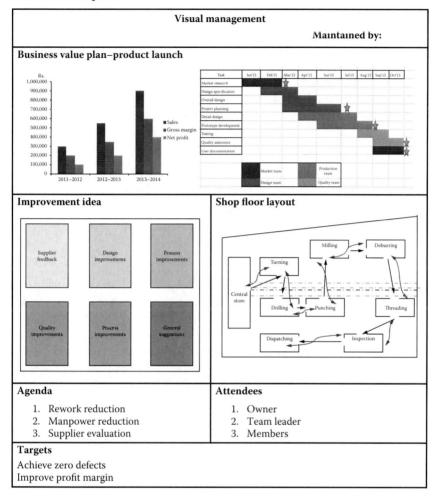

Figure 7.5 A general visual management board displaying manufacturing firm's functional competitiveness.

4. Describe each task in detail. It includes specific execution order, timing sequence, safety considerations and other references.
5. Involve all employees in discussions as the successful deployment of a SOP depends on team effort and action.
6. Set up a system to monitor the SOP regularly. After the implementation of the SOP, it has to be periodically evaluated and updated. During the development of new SOPs, frequent analysis must be done to ensure smooth operation.

7.6.3.1 Applications of SOPs
7.6.3.1.1 Production/operations

- Production process steps, equipment maintenance, inspection pro-cedures, new employee training

7.6.3.1.2 Marketing, sales and customer service

- Authorisation of external communications: press releases, social media, advertising, etc. and preparation of sales quotations, war-ranty, guarantee, refund/exchange policies

7.6.4 An illustrative example

A sample SOP is shown in Table 7.5.

7.7 Cause and effect analysis

7.7.1 What is cause and effect analysis?

During any problem-solving initiative, first a brainstorming session is organised with the individuals associated with the process. During this brainstorming session, the potential causes for the problem are identified. These potential causes to be further analysed to identify the root causes. Hence, as a first step in identifying the root causes, all the potential causes are presented in a cause and effect diagram format (Asaka and Ozeki 1990). For this purpose, all the causes are categorised into different cat-egories like man, machine, material, method, measurement system, etc. and presented in a format as shown in Figure 7.6. This diagram has a cause side and an effect side. The cause and effect diagram is also known as a fishbone diagram or Ishikawa diagram. The Minitab software can be used to prepare a cause and effect diagram. It has provision for 10 catego-ries of causes.

7.7.2 When do you use cause and effect analysis?

Cause and effect analysis is generally used in the following situations:

- To identify key characteristics and key process parameters affecting the output
- To help the group to reach a common understanding of a problem
- To expose gaps in existing knowledge of a problem
- To reduce the incidence of subjective decision-making

Table 7.5 Standard operating procedure

Standard operating procedure (SOP)

Stage	Station name	Model name	Control plan no.	Revision	Date	Final description
1	Piston line	P3-Piston	2			

Process description with photos:

S. no.	Process name	Process description	Safety device	Quality criteria	Reaction plan
11	Turning	Turn the given blank to its specific dimension using turning centre	Shoes, gloves	Ensure proper dimensions with respect to standard part diagram	Redo the operation to achieve the specified dimension
12	Drilling	Drill the machined part to the required dimension using the turning centre	Shoes, gloves	Ensure proper dimensions with respect to standard part diagram	Redo the operation to achieve the specified dimension
13	Bracket assembly	Assemble the brackets in the machined main frame	Shoes, gloves	Ensure proper tightness of main frame with bracket	Retighten to the required torque level
14	Fastener tightening	Assemble the fasteners in the provided slot with respect to brackets	Shoes, gloves, ear plugs	Ensure proper tightness of main frame with bracket	Retighten to the required torque level
15	Heat treatment	Place the assembled components in oven for 45 minutes at 150°C	Shoes, gloves	Test the surface hardness and porosity of the heated part with standard conditions	Test the material properties and reheat the parts
16	Packing	Place the component in the provided bins according to the job number	Shoes, gloves	Ensure proper seating of the parts in the bin	Realign the parts and ensure proper seating

Quality check	Change information
Checked and verified by quality assurance team	
Seal:	**Approvals**
	Proposed by:
	Approved by:
	Released by:

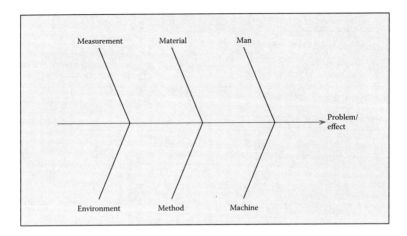

Figure 7.6 General template of cause and effect diagram.

- To help employees to know more about the process and the problems being studied
- To be used as a technical material for further study of the same process or different processes
- To help employees working on the process to investigate the problem and causes associated with it

7.7.3 How do we create cause and effect analysis?

The cause and effect diagram can be created based on the following steps:

- Conduct brainstorming sessions to generate potential causes of the problem
- Group the potential causes in different categories like man, material, machine, method, etc.
- Draw the cause and effect chart using Minitab software
- Check for missing information
- Perform further analysis to identify the root causes out of the potential causes

7.7.4 An illustrative example

A cause and effect diagram is prepared for hardness variation of a component and is presented in Figure 7.7.

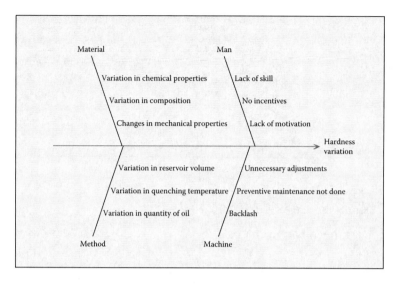

Figure 7.7 An example of cause and effect diagram.

7.8   Pareto analysis

7.8.1   What is Pareto analysis?

A Pareto diagram is a bar chart that helps to prioritise the actions with respect to defects, failures, repairs, customer complaints, etc. This concept was introduced by Wilfred Pareto. The basic principle of Pareto is '80 percent of overall effect is contributed by 20 percent of causes'. Hence, Pareto analysis helps to sort out the 'vital few' from the 'trivial many'.

7.8.2   When do you use Pareto analysis?

When many factors/causes are impacting a problem, a Pareto analysis can be used for the following:

- Find out the most important item/defect
- Identify where to take actions
- Determine ratio of each item to the whole
- Determine degree of improvement after remedial action in some limited area
- Compare improvement in each item/defect before and after the action on the process

Table 7.6 Data on machine breakdown

S. no.	Causes	Frequency
1	Weight variation	430
2	Gum pot leakage	67
3	Top folding problem	1115
4	Label machine packet jamming	124
5	Improper foil feeding	43
6	Packet transfer spring not working	18

7.8.3 How do we create Pareto analysis?

For performing a Pareto analysis, the following steps can be used:

- Define a problem and collect data on the factors that contribute to it. Historical records generally provide sufficient information.
- Arrange the data in descending order and calculate the cumulative percentage.
- Draw the horizontal and vertical axes.
- Prepare a bar graph in the X-axis with descending order of frequency.
- Draw a cumulative percentage graph.

7.8.4 An illustrative example

The data from a tea packaging machine were collected during breakdowns for a period of 1 year, and the summarised data is presented in Table 7.6.

The Pareto can be prepared using the Minitab software by entering the causes in one column and the frequency in another column. The Minitab output of the Pareto analysis is presented in Figure 7.8.

The Pareto diagram shown in Figure 7.8 indicates that the 'top folding problem' is a major cause of breakdowns of the machine. The first two causes together contribute 86% of machine breakdowns.

7.9 Histogram

7.9.1 What is a histogram?

A histogram is a graphical representation of the data which enables us to make quick and objective inferences about the population. A histogram shows the shape or distribution of the data by displaying how often different values occur (i.e. the frequency of occurrence of data points).

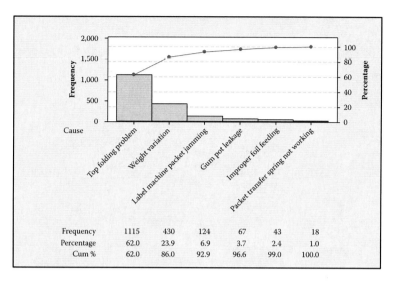

Figure 7.8 Pareto analysis for machine breakdown.

7.9.2 When do you use a histogram?

- A histogram summarises data from a process and graphically presents the frequency distribution. By superimposing specification limits on a histogram, one can judge whether the process is capable of meeting customer requirements.
- Representing and interpreting large amounts of data in tabular form is extremely difficult. By plotting a histogram of the data, the shape, centring and spread of the data can be assessed.
- A histogram shows the relative frequency of occurrence of various data values.
- It also helps to illustrate quickly the underlying distribution of the data.

7.9.3 How do we create a histogram?

The first step in creating a histogram is to collect a sample of 50–100 observations on any characteristic. The histogram can be made only for continuous types of data. After the data collection is complete, the following steps can be used to construct a histogram:

- Find the maximum and minimum values in the sample.
- Prepare class intervals and frequency table.
- Plot the histogram by taking the X-axis as the measured characteristic and the Y-axis as the frequency. The class interval represents

the width of each bar. The height of the bar is proportional to the frequency of the respective class.

- Study the pattern of the plotted histogram and make suitable inference.

7.9.4 An illustrative example

The data shown in Table 7.7 are the chemical process yield on successive days. The Minitab software is utilised for preparing the histogram for these data, and the same is presented in Figure 7.9.

Table 7.7 Data on yield of a chemical process

94.97	87.97	85.21	86.52	92.82	92.06	85.96	88.41	88.20
89.10	91.66	88.97	93.02	88.76	92.57	89.14	87.33	88.91
93.53	87.66	90.08	88.50	89.93	90.04	89.88	90.19	88.59
91.56	91.45	90.23	88.57	90.90	91.18	91.05	91.14	91.20
88.67	88.08	89.46	88.31	90.58	89.75	90.67	89.10	89.40
86.90	89.12	87.00	87.28	90.95	87.64	85.81	88.39	90.91
88.43	92.32	87.39	89.39	88.15	91.35	90.46	94.23	89.95
91.26	89.50	89.24	88.90	92.78	90.09	86.57	90.84	88.72
92.11	86.06	91.06	91.54	86.96	90.07	89.80	88.30	90.45
88.18	86.28	92.17	93.85	88.00	90.14	91.37	90.22	87.53

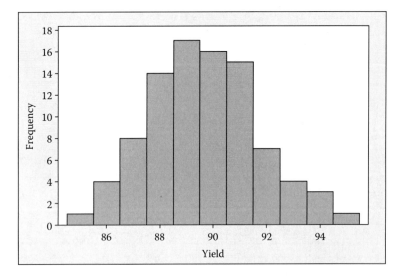

Figure 7.9 Histogram for chemical yield.

The histogram shows a symmetrical shape, with the process centred around the yield value of 90. Also, this does not provide any indication of outliers in the data.

7.10   Scatter diagram and correlation analysis

7.10.1   What is a scatter diagram and correlation analysis?

A scatter diagram is a simple plot of pairs of observations on a graph by taking the independent variable along the X-axis and the dependent variable along the Y-axis. A scatter diagram depicts the picture of a relationship between two variables. If Y increases with X, then X and Y are positively correlated. If Y decreases as X increases, then the two types of data are negatively correlated. If the scatter diagram does not exhibit any identifiable trend, then the variables are said to have no correlation.

The strength of the relationship between variables can be estimated by a term called 'correlation coefficient', generally denoted by 'r'. The value of 'r' measures the magnitude and direction of a linear relationship between two variables X and Y. Its magnitude determines the strength of relationship whereas the sign indicates whether the variables are positively correlated or negatively correlated. The value $r = \pm 1$ implies a perfect linear relationship, and $r = 0$ indicates no relationship between the variables. The correlation coefficient 'r' satisfies the inequality $-1 \leq r \leq 1$.

7.10.2   When do you use a scatter diagram and correlation analysis?

A scatter diagram and correlation analysis are used to study the relationship between two variables. This analysis is very helpful for identifying the root causes of a problem. It is also helpful in determining the optimum operating range for variables in a process.

7.10.3   How do we create a scatter diagram and correlation analysis?

The steps for making a scatter diagram are as follows:

- Collect 50–100 paired samples of data believed to be related.
- Draw the horizontal and vertical axes of the diagram and label the axes.
- The variable that is being investigated as the possible cause/input is usually on the horizontal axis (X-axis), and the effect/result/output variable is usually on the vertical axis (Y-axis).
- Plot the data on the diagram, and study the pattern of the data.

The steps for calculating the correlation coefficient are as follows:

- Collect the (X, Y) data in pairs.
- Enter the data in two different columns of Minitab software. Use the Minitab command of '*Stat > Basic Statistics > Correlation*' to obtain the correlation coefficient.

7.10.4　An illustrative example

Consider the data in Table 7.8 of temperature (X) and strength (Y). These data are used for making a scatter diagram and calculating a correlation coefficient. The scatter plot is prepared for these data by using the Minitab software and is presented in Figure 7.10.

Table 7.8 Data on temperature (X) and strength (Y)

X	Y	X	Y	X	Y	X	Y	X	Y
177	208	207	295	200	277	199	273	211	308
179	214	209	295	201	298	188	277	210	295
188	236	209	298	196	280	209	286	204	295
183	242	220	302	203	289	199	292	206	292
194	248	209	298	178	217	207	305	198	295
191	252	221	302	179	214	206	298	207	280
200	267	213	292	186	233	209	305	202	289
194	264	216	292	183	242	210	302	205	289
205	286	203	292	193	245	215	305	178	211
204	289	205	292	181	248	206	302	180	220

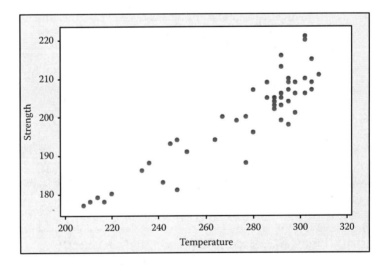

Figure 7.10 Scatter diagram for temperature vs. strength.

The scatter diagram gives an indication that as the temperature increases, the strength also increases. The correlation coefficient was estimated using the Minitab software, and the value was obtained as $r = 0.903$. This shows a strong positive relationship between the X and Y variables.

7.11   Control charts

7.11.1   What are control charts?

The concept of a control chart was introduced by Walter A. Shewhart in the 1920s. The control chart is a line graph used to assess the stability of a process. It is based on the principle of normal distribution. The control chart has three lines drawn on it, namely the central line (CL), upper control limit (UCL) and lower control limit (LCL). The centre line is the average, and the upper and lower control limits are calculated based on 'average ± 3 standard deviation' basis. The points are plotted on this chart, and the pattern of behaviour of the graph is studied to infer about the stability of the process (Montgomery 2009). If any point falls beyond the control limits, it can be a signal for special causes and will need further investigation. Figure 7.11 presents a typical sketch of a control chart.

Depending on the category of data, there are different types of control charts available. The control charts used for variable data are known as variable control charts, and the controls used for attribute data are known as attribute control charts. Under variable control charts, depending on the sample size, different types of charts, namely X-bar and R chart, X-bar and S chart etc. are used. Similarly, in the attribute category we have p-chart, np-chart, c-chart and u-chart (Table 7.9).

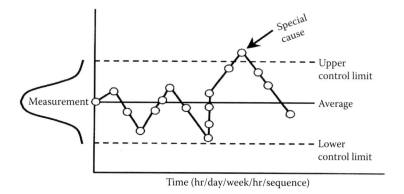

Figure 7.11 A typical sketch of control chart.

Table 7.9 Summary of different types of control charts

Data type	Name of chart
Continuous with subgroup size ≤10	$\bar{X}$ –R chart (X-bar and R chart)
Continuous with subgroup size >10	$\bar{X}$ –S chart (X-bar and S chart)
Discrete, defective type, sample size not constant	P-chart
Discrete, defective type, sample size constant	NP-chart
Discrete, defect type, sample size not constant	U-chart
Discrete, defect type, sample size constant	C-chart

7.11.2 When do you use control charts?

The control charts are used for the following purposes:

- To study the stability of the process.
- To analyse the processes to identify the presence of assignable causes or special causes so that actions can be initiated for improvement.
- To monitor the process over a period of time.

7.11.3 How do we create a control chart?

The general steps for creating a control chart are as follows:

- Collect chronological data from the process.
- Depending on the category of data, decide on the type of control chart.
- Use the Minitab software to plot the chart.
- Study the chart for assignable causes.
- Take action to understand the assignable causes and try to reduce or eliminate them from the process.

7.11.4 An illustrative example

Data on board thickness (in inches) are given in Table 7.10 for 15 samples of three boards each.

As these data are collected in subgroups, and the subgroup size is three, an X-bar R chart can be used. The X-bar R chart is plotted with the help of Minitab software. The output of the Minitab software is presented in Figure 7.12.

In this X-bar R chart, all the plotted points are falling within the control limits without any specific pattern. Hence we can conclude that there are no assignable causes present in the process.

Table 7.10 Data on thickness

Sample no.	Subgroup		
1	0.0629	0.0636	0.0640
2	0.0630	0.0631	0.0622
3	0.0628	0.0631	0.0633
4	0.0634	0.0630	0.0631
5	0.0619	0.0628	0.0630
6	0.0613	0.0629	0.0634
7	0.0630	0.0639	0.0625
8	0.0628	0.0627	0.0622
9	0.0623	0.0626	0.0633
10	0.0631	0.0631	0.0633
11	0.0635	0.0630	0.0638
12	0.0623	0.0630	0.0630
13	0.0635	0.0631	0.0630
14	0.0645	0.0640	0.0631
15	0.0622	0.0644	0.0632

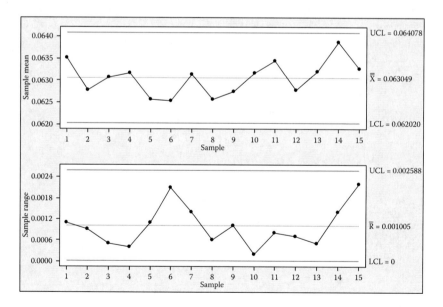

Figure 7.12 X-bar R chart for thickness.

7.12   Run charts

7.12.1   What are run charts?

A run chart is used to monitor the behaviour of a variable over time for a process or system. Run charts graphically display cycles, trends, shifts or non-random patterns in behaviour over time. A trend is an unusually long series of consecutive increases or decreases in the data. Run charts can help identify problems and the time when a problem occurred, or monitor progress when solutions are implemented.

7.12.1.1   Questions to ask about a run chart
1. Is the average line where it should be to meet customer requirements?
2. Is there a significant trend or pattern that should be investigated?

7.12.1.2   Two ways to misinterpret run charts
1. You conclude that some trend or cycle exists, when in fact you are just seeing normal process variation (and **every** process will show some variation).
2. You do not recognize a trend or cycle when it **does** exist.

Both of these mistakes are common, but people are generally less aware that they are making the first type, and are tampering with a process which is really behaving normally. To avoid mistakes, use the following rules of thumb for a run chart interpretation:

1. Look at data for a long enough period of time, so that a 'usual' range of variation is evident.
2. Is the recent data within the usual range of variation?
3. Is there a daily pattern? Weekly? Monthly? Yearly?

7.12.2   When do you use run charts?

Run charts are used for the following purposes:

- To understand the stability of a process over time.
- To understand patterns and trends in a process.
- To evaluate the success of improvement efforts in an objective and visual way.

Typical applications include charting process and product defects over time, logging computer or system downtime, assessing process yield over time, etc.

7.12.3 How do we construct a run chart?

The following are useful in the construction of a run chart:

- Determine the purpose of the chart and the data to be monitored, collected and analysed. Select the time interval (minute, hour, day, month, etc.).
- Collect the data that will be plotted.
- Plot time on the horizontal axis (typically located at the bottom).
- Label the vertical axis and plot the data collected in step 2 on the vertical axis (typically located on the left-hand side).
- Title the chart. Indicating which direction is better with an up or down arrow may be helpful.

7.12.4 An illustrative example

Figure 7.13 shows a run chart for the number of software discrepancies per modified line of code detected by a customer of a software production company.

The different dots represent data for different revisions of some major source of source code. It is important to note that there are fluctuations in the output, and it is hard to say if the process is unstable from a small sample size as shown in Figure 7.13.

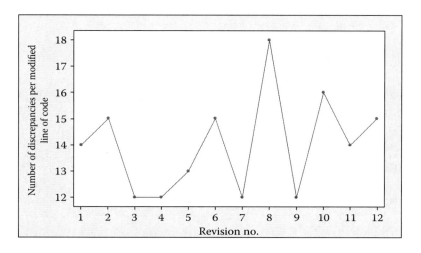

Figure 7.13 Example of a run chart.

7.13 Failure Modes and Effects Analysis

7.13.1 What is a Failure Mode and Effects Analysis?

Failure modes and effects analysis (FMEA) is a step-by-step approach for identifying all possible failures in a design, a manufacturing or service process, or a system or assembly. 'Failure modes' means the ways, or modes, in which something might fail. Failures are any errors or defects, especially ones that affect the customer, and can be potential or actual. 'Effects analysis' refers to studying the consequences of those failures. Failures are prioritised according to how serious their consequences are, how frequently they occur and how easily they can be detected. The purpose of the FMEA is to take actions to eliminate or reduce failures, starting with the highest priority ones.

FMEA includes review of the following:

- Steps in the process
- Failure modes (what could go wrong?)
- Failure causes (why would the failures happen?)
- Failure effects (what would be the consequences of each failure?)

7.13.2 When do we use an FMEA?

FMEA has been proved to be useful when

- One has to evaluate processes for possible failures and to prevent them by correcting the processes proactively rather than reacting after failures have occurred.
- One has to evaluate a new process prior to implementation and in assessing the impact of a proposed change to an existing process.
- One has to design a new product or process or service (i.e. early design stages) to ensure the design is robust (i.e. less sensitive to failures in the user's environment).

7.13.3 How do we create an FMEA worksheet?

The basic steps for conducting an FMEA are

Step 1: Study the product/process/service/system/assembly to be analysed.

Step 2: Assemble a cross-functional team of people with diverse knowledge about the process, product or service and customer needs.

Step 3: Identify the scope of your FMEA exercise. Identify the functions of your scope. Ask, 'What is the purpose of this system, design, process or service? What do our customers expect it to do?'

Step 4: For each function, identify all the ways that failure could happen. Brainstorm the possible range of failure modes.

Step 5: List all the potential consequences of each failure mode.

Step 6: Assign severity scores (S) for each failure mode. Severity is usually rated on a scale from 1 to 10, where 1 is insignificant and 10 is catastrophic. If a failure mode has more than one effect, write on the FMEA table only the highest severity rating for that failure mode.

Step 7: Identify the causes of each failure mode (you may use cause and effect analysis here).

Step 8: For each cause, determine the occurrence rating (O). This rating estimates the probability of failure occurring for that reason during the lifetime of your scope. Occurrence is usually rated on a scale from 1 to 10, where 1 is extremely unlikely and 10 is inevitable. On the FMEA table, list the occurrence rating for each cause.

Step 9: Identify control plans to detect the failure modes.

Step 10: For each control, determine the detection rating (D). This rating estimates how well the controls can detect either the cause or its failure mode after they have happened but before the customer is affected. Detection is usually rated on a scale from 1 to 10, where 1 means the control is absolutely certain to detect the problem and 10 means the control is certain not to detect the problem (or no control exists). On the FMEA table, list the detection rating for each cause.

Step 11: Calculate the risk priority number (RPN), which is the product of severity, occurrence and detection. These numbers provide guidance for ranking potential failures in the order they should be addressed.

Step 12: Prioritise the failure modes and causes based on the RPN.

Step 13: Determine the actions to be taken. These actions may be design or process or system changes to lower severity or occurrence. They may be additional controls to improve detection. Also note who is responsible for the actions and target completion dates.

Step 14: Recalculate the RPN values based on the action plans.

7.13.4   *An illustrative example*

Consider the process for dispensing fuel at a self-service gas station. There are a number of steps or sub-processes involved in this case starting from finding a gas pump all the way through to remitting payment. In order to make things easier for the reader, we carry out the FMEA exercise on three sub-processes – find gas pump, initialise pump and dispense fuel. The FMEA worksheet is shown in Table 7.11. The higher the RPN, the more critical a failure mode can be, and we should take remedial actions to minimise the impact of those failure modes. In the given example, the most critical failure mode is when auto shut-off fails. We need to

Table 7.11 FMEA worksheet

Process steps	Potential failure modes	Severity (S)	Occurrence (O)	Detection (D)	RPN = S*O*D	Revised RPN (after actions to improve the process)
Find gas pump	All pumps are busy	1	6	1	6	
	Gas cap on other side of the vehicle	4	4	10	80	
Initialise pump	Pump won't reset	4	1	1	4	
	Pump won't read credit card	8	1	1	8	
Dispense fuel	Auto shut-off fails	10	2	10	200	

investigate the causes for this failure mode and develop an action plan to prevent the occurrence of this failure mode. Once the remedial actions are implemented, one may recalculate the RPN values and see if the remedial actions are effective or not.

7.14 Voice of the Customer analysis

7.14.1 What is a VOC analysis?

The 'voice of the customer' (VOC) is a process used to capture the requirements/feedback from the customer (internal or external) to provide the customer with the best-in-class service/product quality. This process is all about being proactive and constantly innovative to capture the changing requirements of the customers with time. The VOC is the term used to describe the stated and unstated needs or requirements of the customer. The VOC can be captured in a variety of ways: direct discussion or interviews, surveys, focus groups, customer specifications, observation, warranty data, field reports, complaint logs, etc. Lean Six Sigma (LSS) deployment aligned to VOC enables companies to constantly improve their overall competitive value proposition, increase market share and improve profitability (Gopalakrishnan 2010).

There are two basic types of VOC data. The first is called reactive data and can be found as customer complaints, compliments, feedback, hotline data, product returns and/or warranty claims. Perhaps to no surprise, this data type is usually negative, and however difficult to hear, it typically represents significant improvement opportunities. For example, it is likely that a customer complaint occurs after the person experiences poor service or receives a bad product. Other dissatisfied customers may not announce a complaint and may just immediately switch to a competitor. The second data type is called proactive data and can be collected from customer interviews, surveys, focus groups, observations and/or test customers. This data type can also identify important improvement opportunities.

7.14.2   When do we use VOC analysis?

VOC analysis is used

- To identify potential improvement opportunities and the associated priorities from a customer viewpoint
- To focus on those items that are most valuable to current and future customers
- To develop the critical to quality (CTQ) characteristics which are critical to your customers

7.14.3   How do I perform VOC analysis?

The following steps can be followed for performing VOC analysis:

Step 1: Identify your customers and what you need to know about their needs – you can have internal and external customers in any organisation. You work primarily with the people involved with the next process step, but you check to ensure that their needs are consistent with the needs of the final customer. Different customers have different needs and priorities. When you carry out an LSS project, you need to understand how the project is aligned with customer satisfaction.

Step 2: Determine the tools and techniques you will use to gain feedback from your target customers. Surveys, focus groups, and one-on-one interviews are popular tools used to gather the information. One of the most frequent mistakes at this stage is over-reliance on surveys; although interviews and focus groups are more labour intensive than surveys, they are generally much more effective in gathering detailed qualitative information that can help lead to improvement.

Step 3: Analyse the data you collect, and summarise the information in a meaningful way.

Step 4: Develop a schedule to reassess your customers on an ongoing basis to ensure your product or service offerings meet or exceed their requirements.

7.14.4   An illustrative example

In this example, we consider a scenario where the customer would like to apply for a mortgage from a local bank. The VOC in this particular case (from surveys and focus group interviews) identifies friendly staff, knowledgeable staff, speed of making decisions and accuracy of information provided to customers. However, from the bank's perspective, the customers' voice is not so clear and therefore, it is essential to go back to customers and understand their voice better. Further dialogues with customers indicated that friendly staff implies willingness to help customers with their questions, being polite to them and so on. Knowledgeable staff implies the mortgage application process from start to finish, understanding the finance issues around the mortgage so that customers will get the best mortgage deal, knowing the market, having a good understanding of the competitors and their products and offers, etc. The speed of making decisions is dependent on the application form itself (length of the form), manual vs electronic, checking all the required documents of customers for making decisions and so on. Finally, the accuracy of information implies executing things right first time, giving the correct information right first time, etc. This example clearly illustrates the point that the VOC at the outset was quite obscure; however, simple tools such as surveys and focus group interviews assist the bank to understand what the needs and expectations of customers are from a mortgage application process.

7.15   CTQs and a CTQ tree

7.15.1   What are CTQs?

The CTQ concept is an essential part of LSS projects. CTQ characteristics (CTQs) are developed in order to fulfil the needs of valuable customers. Customer satisfaction is a primary factor in the development of CTQ parameters. CTQs analyse the characteristics of the service or product that are termed by both the internal and external customer (Gopalakrishnan 2012).

A CTQ tree reflects what the customers of your process cite as absolutely essential to success. This helps clarify what constitutes a defect in the process. Use a CTQ tree when you want to investigate your customer base by moving from general needs to more specific ones (https://www.mindtools.com/pages/article/ctq-trees.htm).

7.15.2 Why do we need CTQs?

CTQs

- Link customer needs gathered from VOC data collection efforts with specific and measurable characteristics
- Enable us to focus on certain quality characteristics which are critical to our customers, especially when they are broad, vague and complex

7.15.3 How do we construct a CTQ tree?

The following steps can be used to construct a CTQ tree:

Step 1: Identify critical needs that your product or service has to meet – During this first step, you're essentially asking, 'What is critical for this product or service?' It is best to define these needs in broad terms; this will help ensure that you don't miss anything important in the next steps.

Step 2: Identify quality drivers – What are the specific quality drivers that meet the needs of your customers? Tools such as Kano analysis would be extremely useful to understand the quality drivers and even the importance of such drivers.

Step 3: Identify performance requirements – You need to identify the minimum performance requirements that you must satisfy for each quality driver, in order to actually provide a quality product. Once you've completed a CTQ tree for each critical need, you'll have a list of measurable requirements that you must meet to deliver a high-quality product.

7.15.4 An illustrative example

In this example, we consider a simple example of a store that sells baby clothing. After talking to a number of potential customers, one of the critical needs highlighted by customers is 'good customer service'. However, the term 'good customer service' may mean different things to different customers, and it is not at all easy to understand what constitutes a good service unless we develop a CTQ tree to understand the quality drivers and the performance requirements associated with these drivers. Figure 7.14 shows a CTQ tree for this example.

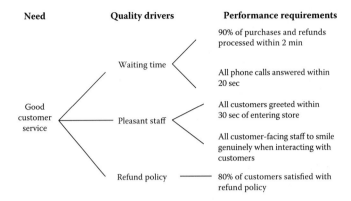

Figure 7.14 Example of a CTQ tree.

7.16   Project charter

7.16.1   What is a project charter?

The project charter is an official, basic document that outlines an LSS improvement project. It takes place in the define phase of DMAIC. However, it can be periodically reviewed, refined and revised throughout the project. The elements of a project charter can vary, but they generally include the business case, problem statement, goal statement, team members/roles and constraints/boundaries and project scope. If done correctly, a project charter will be your go-to guide when it comes to keeping teams, stakeholders and leaders on track through a predetermined guide that must be available to everyone working on the project.

7.16.2   Why do we need a project charter?

A project charter

- Provides a reference point that the team can continually refer back to in order to keep direction and scope
- Gives the LSS project champion and the project team authority to use organisational resources for the project and forms the basis of communication with stakeholders
- Keeps the project from drifting off course by focusing on predetermined goals and objectives
- Indicates why a project has priority over other projects
- Describes the gap between the current state and the desired state, especially when we refer to any process improvement projects

7.16.3 How do we construct a project charter?

The following steps can be used in the development of a project charter:

Step 1: Define the business case. Why is the project important, and why now?

Step 2: Define the problem statement. Quantify the problem and its impact on the business and its customers. The baseline of the problem must be understood by everyone, and this might require data collection.

Step 3: Define the project scope. What are the boundaries of the project, what is in scope, and (sometimes even more important) what is out of scope?

Step 4: Define the goal(s) of the project. What are the tangible performance improvements that the team has as a target?

Step 5: Define the roles and responsibilities of each team member of the project. Also, determine the resources required in carrying out the project.

Step 6: Define the key deliverables of the project along with the milestones.

Step 7: Review the project charter with the project champion and gain approval. If any changes are required, it is essential to make changes and get them approved by the champion.

7.16.4 An illustrative example

A company uses infrared (IR) lasers for cutting materials such as glass, solar panels, etc. The cross-sectional shape and propagation properties of the laser beam have a massive impact on the quality of the cut that the laser makes in the material being processed. The laser's IR mode is set by a laser systems engineer, and the mode is measured using a beam profiler and is brought into specification by correct selection and positioning of four lenses. The time taken to set the mode varies greatly from 2 hours to nearly 40 hours. This has been an ongoing issue for a long period of time, and it has received senior management's attention as a priority.

The project was limited to improving the time taken to set the mode through

- Reducing the variation of the raw mode
- Improving the predicted lens choice and position
- Ensuring the relevance of the choice of lenses available

The goal of the project is to reduce the variation in the time taken to set the mode. In the first phase of improvement, the target time for setting the IR mode is 6.5 hours ± 1 hour. Anything which is outside the 5.5 and 7.5 hours is viewed as a defect. The estimated savings of the project were close to £100,000 per year.

The team members for the project included

- Two Six Sigma green belts
- Production manager
- Project champion
- Project sponsor (director of manufacturing)
- Laser systems engineer

7.17   Hypothesis tests

7.17.1   What are hypothesis tests?

There are many statistical problems in which we are not directly concerned with the actual value of a parameter of a distribution; instead we want to know whether its value exceeds a given value, or is less than a given value, or falls into a given interval and so on. Hence, instead of estimating the value of a population parameter, a decision about the correctness of a statement concerning the parameter is tested (Kiemele et al. 2000). Thus, a hypothesis about the parameter is tested. This hypothesis can be based on any population parameter, viz, average, variance, proportion, etc.

Hence, a test of hypothesis is a test of the validity of the assertion, and the validity is determined by an analysis of the data. Thus, a test of hypothesis is a set of rules on the basis of which a decision is to be taken on whether to accept the hypothesis or reject it. Two different hypotheses are defined during this test, namely null hypothesis and alternate hypothesis. A hypothesis which needs to be validated by data is known as a 'null hypothesis', denoted by H_0. An alternate hypothesis, denoted by H_1, is a hypothesis that is accepted when the test of hypothesis leads to rejection of the null hypothesis (Montgomery and Runger 2007).

The different types of hypothesis tests available include the tests for averages, tests for variation, tests for proportion, tests for independence, etc. Some of the commonly used tests are summarised in Table 7.12.

7.17.2   When do you use hypothesis tests?

Hypothesis tests are used for the following situations:

- To prove or disprove any assumptions about a process
- To make comparison of the processes before and after improvement
- To make inferences about unknown population parameters

Table 7.12 Different types of hypothesis tests

Situation	Test name
Comparing an average with a constant	One sample t-test
Comparing averages of two populations	Two sample t-test
Comparing averages of multiple populations	Analysis of variance (ANOVA)
Comparing variance with a constant	Chi-square test
Comparing variances of two populations	F-test
Comparing variances of multiple populations	Bartlett's test
Testing the independence of attributes	Chi-square test

7.17.3 *How do we perform hypothesis tests?*

- Convert the practical problem to a statistical problem by defining the null hypothesis (H_0) and alternative hypothesis (H_1).
- Identify the test to be performed.
- Using Minitab software, obtain the p-value of the test.
- If the p-value is <0.05, reject H_0 at 5% level of significance. If the p-value is <0.01, reject H_0 at 1% level of significance.

7.17.4 *An illustrative example*

Example 1

The shelf life of a photographic film is of interest to the manufacturer. The manufacturer observes the following shelf life for eight units (in days) chosen at random from the current production: 108, 128, 134, 163, 124, 159, 116, 134. Is there any evidence that the mean shelf life is greater than 125 days?

Solution

In this case the null hypothesis is H_0: $\mu = 125$, and the alternative hypothesis is H_1: $\mu > 125$.

The test to be used is one sample t-test. This analysis is performed using Minitab software (Minitab version 16), and the output is given as follows:

Variable	N	Mean	St. Dev	Std. Error	T	p-value
Time	8	133.25	19.26	6.81	1.21	0.133

Since the p-value is 0.133, greater than 0.05, we accept the null hypothesis. That means there is no evidence that the mean shelf life is greater than 125 days.

Example 2

A new filtering device is installed in a chemical unit. Before and after its installation, a random sample yielded the following information about the percentage of impurity: $\bar{x}_1 = 12.5$, $s_1^2 = 101.17$, $n_1 = 8$, $\bar{x}_2 = 10.2$, $s_2^2 = 94.73$, $n_2 = 9$. Can you conclude that the two variances are equal?

Solution

In this case, the null hypothesis is H_0: $\sigma_1^2 = \sigma_2^2$, and the alternate hypothesis is H_1: $\sigma_1^2 \neq \sigma_2^2$.

The test to be used is the F-test as we compare two variances. This analysis is performed with Minitab software, and the output is given as follows:

Test

Method	DF1	DF2	F-statistic	p-value
F-test	7	8	1.07	0.918

Since the p-value is 0.918, we conclude that there is no significant difference between the variances of the processes.

7.18   Regression analysis

7.18.1   What is regression analysis?

The technique of regression analysis helps in estimating the relationship in mathematical form and measures the strength of that relationship. In regression analysis, a functional relationship of the form $Y = f(X)$ between two variables is identified.

The simplest form of regression analysis is the study of the linear relationship between X and Y of the form $Y = a + b\,X$, where a and b are constants to be calculated from the data. This type of regression is generally known as *simple linear regression*. The coefficients a and b are estimated based on the principle of maximum likelihood (Draper and Smith 2003).

A generalisation to this simple linear regression is the multiple linear regression of the form $Y = a_0 + a_1 X_1 + a_2 X_2 + a_3 X_3 + \ldots + a_k X_k$, where $a_1, a_2, \ldots, a_k$ are constants.

7.18.2   When do you use regression analysis?

The regression analysis is used to estimate the relationship between variables (Montgomery and Peck 1982). It also helps to identify a model between the output characteristics and input variables. It helps to study the cause–effect relationship in a problem-solving scenario. Once a regression model is developed, one can predict the output for a given value of input or given values of inputs.

7.18.3 How do we perform simple linear regression analysis?

The following steps are used to create a simple linear regression analysis:

- Collect the data on X and Y
- Estimate the constants a and b of the equation $Y = a + bX$ (using Minitab software)

7.18.4 An illustrative example

Consider the data on temperature and strength provided in Table 7.8 (see Section 7.10.4). These data can be used for modelling a simple linear regression equation of the form $Y = a + b\,X$. For this purpose, first the data are entered in two different columns of Minitab software. After performing the analysis, the output is presented in Figure 7.15.

Here, we obtain an equation, *Strength* = –187.2 + 2.314 *Temperature*. This equation can be used for predicting the *strength* values for any value of *temperature*.

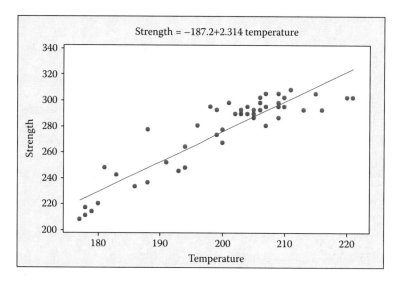

Figure 7.15 Regression analysis.

7.19   Kanban system

7.19.1   What is Kanban?

Kanban is a visual signal that is used to initiate an action. The word *Kanban* in Japanese means 'card'. A Kanban system is a work scheduling system that maximises the productivity of a team by idle time reduction. Kanban starts with a customer's order and follows production downstream. Kanban is referred to as a 'pull' system as it works based on customer order. A part is only manufactured (or ordered) if it contains a Kanban card. Kanban is used in the improve phase of Six Sigma methodology.

Kanban is used in a pull production system where customer orders form the basis for product manufacture. The system is simple and can be readily used. Kanban delivers the exact quantity of material to the right location based on the requirement. A pull system visualises the manufacturing process from the viewpoint of the finished part. The production controller works based on orders representing the firm's customer requirements.

7.19.1.1   Rules for Kanban

- Kanban authorises production and withdrawal of parts.
- The containers/bins must always be accompanied by a Kanban.
- The order of receipt of Kanbans sets the priority order for part manufacture.
- Kanban specifies standard production quantity.
- Kanban does not allow the dispatch of defective parts.

7.19.2   When do you use Kanban?

- To establish an upper threshold to work in progress inventory
- To identify opportunities for improving efficiency
- To improve flexibility
- To focus on continuous delivery
- To increase efficiency

7.19.3   How do we create a Kanban?

1. Map the value stream
2. Define the starting and ending points for the Kanban system
3. Agree with initial WIP limits and change procedures
4. Develop a Kanban card
5. Initiate deployment

7.19.3.1   Calculation of number of Kanbans

The number of Kanbans is calculated using Equation 7.1 (Haslett and Osborne 2000):

Number of Kanbans = (average demand during lead time + safety stock)/container quantity

$$N = (dL + S) / C \qquad (7.1)$$

where
- N = Number of Kanbans
- d = Average demand per hour
- L = Lead time in hours
- S = Safety stock
- C = Container quantity

7.19.4 An illustrative example

A firm produces 100 components per day working at 8 hours. Calculate the Kanban size and number of Kanban cards required if the replenishment time is 20 hours and the lot size is 60 parts.

The Kanban quantity and number of Kanban cards required are calculated using Equations 7.2 and 7.3:

$$\text{Kanban quantity} = \frac{(\text{Daily production rate} \times \text{Replenishment time})}{(\text{Available time})}$$

$$= \frac{100 \text{ components} \times 20\,\text{h}}{(8\,\text{h})} = 250 \text{ parts} \qquad (7.2)$$

$$\text{Number of Kanban cards} = (\text{Kanban quantity}) / (\text{Lot size})$$

$$= (250) / (60) = 4 \text{ Cards} \qquad (7.3)$$

A sample Kanban card is shown in Figure 7.16.

Part description			Part number		
M-ring			4123339		
Quantity	250	Lead time	1 week	Order date	06/03/2015
Supplier		M/s XYZ rings	Due date	13/03/2015	
Contact		Mr. Kumar		Card 1 of 4	
				Location	Rack 112

Figure 7.16 Sample Kanban card.

7.20   Poka-Yoke (mistake proofing)

7.20.1   What is Poka-Yoke?

Poka-Yoke is a Japanese term that implies 'mistake proofing' (Shingo 1986). A Poka-Yoke is a mechanism that aids an equipment operator to avoid mistakes at the first instance. The goal of Poka-Yoke is to eliminate product defects by prevention, rectification and correction on human errors. Shigeo Shingo was instrumental in conceptualizing Poka-Yoke to achieve zero defects and eliminate inspection. Poka-Yoke is a meticulous approach to preventing errors by analysing their root causes. A Poka-Yoke is a device that either prevents or detects abnormalities, which may impact product quality or employees' health and safety. Categories of Poka-Yoke include prevention and detection. Prevention is engineering the process in such a way that the occurrence of errors or mistakes could be prevented. In detection, the user is alerted by a signal on occurrence of an error to rectify the mistake.

7.20.2   When do you use Poka-Yoke?

The rationale behind Poka-Yoke is to focus on the intelligence of workers by avoiding repetitive activities. Poka-Yoke allows the workers to concentrate on creative value-adding activities (Shingo 1986). Poka-Yoke involves techniques that make it impossible to commit mistakes. These techniques enable driving out product/process defects and improve quality and reliability. The use of simple Poka-Yoke in product and process design can eliminate both mechanical and human errors.

There are different types of Poka-Yoke:

- Contact
- Fixed value
- Motion stop

Contact Poka-Yoke: Contact-type Poka-Yoke devices include physical shapes that are used to prevent the use of wrong components, pins which must mate with holes from prior operations. The mechanism is that this Poka-Yoke makes physical contact with the product and signals the errors.

Fixed value Poka-Yoke: This type uses physical and visual methods to ensure the availability of components in the right quantities.

Motion stop Poka-Yoke: These devices ensure that adequate numbers of steps as well as the sequence of steps have been taken.

7.20.3   How do we perform Poka-Yoke?

The step-by-step procedure involved in applying Poka-Yoke concepts (Stewart and Melnyk 2000) is discussed as follows:

1. By using Pareto analysis, identify the process or operation where the occurrence of error is high.
2. Using 5 Whys analysis or root cause analysis (refer to Section 7.20), analyse and explore the ways of process failure.
3. Select the appropriate Poka-Yoke approach.
4. Develop a comprehensive approach of Poka-Yoke.
5. Determine whether a contact type, fixed value or motion stop technique is most appropriate in addressing the error.
6. Perform trials and explore its workability.
7. Provide suitable training to the operator and review his or her performance to ensure success.

7.20.4   Illustrative examples

7.20.4.1   Illustrative examples of Poka-Yoke

- An interference pin that prevents workers from installing a part in the wrong way (Dvorak 1998)
- Usage of push and lift symbols in beverage vending machines (Mahapatra and Mohanty 2007)
- Mobile phone SIM card which can only be fitted one way in the mobile (Lockton et al. 2010)

7.21   Root cause analysis or 5 Whys analysis

7.21.1   What is root cause analysis?

Root cause analysis or 5 Whys analysis is a simple but powerful tool for quickly uncovering the root of a problem, so that you can deal with it once and for all. Sakichi Toyoda, one of the fathers of the Japanese industrial revolution, developed this problem-solving tool in the 1930s, but it became more popular in the 1970s, and Toyota still uses it to solve quality- and process-related problems today.

Root cause analysis is most effective when the answers come from people who have hands-on experience of the process being examined. It is remarkably simple: when a problem occurs, you uncover its nature and source by asking 'why' no fewer than five times (http://asq.org/service/body-of-knowledge/tools-5whys).

7.21.2   When do you use a root cause analysis?

A root cause analysis is used

- When you need to determine the root cause of a problem in problem-solving scenarios
- When problems involve human factors or interactions
- In conjunction with brainstorming and cause and effect analysis

7.21.3   How do we carry out a root cause analysis?

Step 1: Identify the key stakeholders who should be involved in the process.

Step 2: Determine the problem that you want to analyse (e.g. why are sales down?). Clearly identify and document the problem or question.

Step 3: Identify appropriate responses to the question (e.g. people don't like our products).

Step 4: Ask 'why' you received the answer to the previous question (e.g. why don't people like our products?).

Step 5: Continue these steps until you reach what could be deemed the root cause. This will generally not require asking 'why' more than five times.

7.21.4   An illustrative example

The quality assurance manager of a company working in a highly government-regulated industry has recognised that most of his staff members spend a large proportion of their time reviewing and correcting documentation. In a staff meeting, a root cause analysis was performed to determine the root cause. Figure 7.17 shows an illustrative example of a root cause analysis. The root cause of the above problem was that there were no standard procedures put in place for the preparation of documents. Moreover, there has been a common problem with lack of training received by the staff members in this particular case.

7.22   Design of Experiments

7.22.1   What is design of experiments?

Design of experiments (DoE) is a powerful technique that can be used for process optimisation scenarios. DoE allows for multiple input factors to be manipulated determining their effect on a desired output (response). In the context of LSS, the response can be a CTQ. By manipulating multiple inputs at the same time, DoE can identify important interactions that may

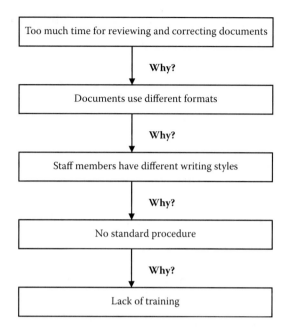

Figure 7.17 An illustrative example of root cause analysis.

be missed when experimenting with one factor at a time (Montgomery 1991). All possible combinations can be investigated (full factorial) or only a portion of the possible combinations (fractional factorial).

Many of the current statistical approaches to designed experiments originate from the work of Ronald A. Fisher in the early part of the 20th century. A well-performed DoE may provide answers to questions such as

- What are the key factors or process variables or inputs in a process?
- At what settings would the process deliver acceptable performance?
- What are the key, main and interaction effects in the process?
- What settings would bring about less variation in the output or response or CTQ?

7.22.2 When do you use DoE?

DoE is most useful when you have to

- Determine the best combination of factor or process parameter settings that produce the best output (s) at lowest cost.
- Identify and quantify the factors or process parameters that have the biggest impact on the output.

- Identify the factors or process parameters that do not have any impact on quality or customer satisfaction so that their levels can be set at the most economical level.
- Reduce the time and costs associated with experiments especially when you have to study a large number of variables with minimum budget at hand.
- Understand the factors or process parameters which influence the average of the response or CTQ and the variation in response or CTQ.
- Build a mathematical model relating the response or output to the key input process variables or factors.

7.22.3   How do we perform a design of experiments?

Step 1: Define the problem in your process.

Step 2: Determine and agree the objectives of the experiment.

Step 3: Identify the response or CTQ which is relevant to the problem at hand.

Step 4: Brainstorm all the potential process variables or factors which could potentially impact the CTQ.

Step 5: Identify the process parameters and their levels for the experiment.

Step 6: Choose a strategy for experimentation depending on a set of criteria (this may include complexity of the problem, degree of optimisation required, objective of the experiment, costs implications, degree of statistical robustness, etc.).

Step 7: Design the experimental layout using a software system such as Minitab.

Step 8: Conduct the experiment and collect data. Data can be directly fed into Minitab.

Step 9: Analyse the data and interpret the results.

Step 10: Determine the optimal settings for the process under investigation.

Step 11: Run confirmatory experiments to verify that the results are sound and valid.

Step 12: Capture the key lessons learned – what has gone right and what has gone wrong?

7.22.4   An illustrative example

A chemical engineer in an SME was keen to improve the yield of a chemical process. Further to a brainstorming session with the production supervisor, production engineers, quality engineer, operators and a process improvement engineer in the company, three potential process parameters were thought to affect the yield. The list of parameters and their levels are shown in Table 7.13.

The team was interested to understand the effects of all three process parameters and the interactions between them (if any exist). It was important to analyse all the two-factor interactions, and therefore a 2^3 full factorial design (i.e. eight experimental runs) was chosen. Each trial condition was replicated three times in order to obtain an accurate estimate of experimental error (or error variance).

In order to design the experimental layout, Minitab software system (version 16) was used. Table 7.14 shows the actual settings of the process parameters with the response values (i.e. yield) recorded at each experimental trial condition.

Figure 7.18 illustrates the Pareto plot of effects. The graph shows that main effects T (temperature) and R (reaction time) and interaction between pressure (P) and reaction time (R) are significant at 5% significance level (i.e. the risk). It is quite interesting to note that pressure (P) on its own has no significant impact on the process yield. It is important to analyse the interaction between P and R to determine the best settings for optimising the chemical process yield.

Figure 7.19 indicates that there exists a strong interaction between pressure and reaction time. It is clear that the effect of the reaction time at different levels of pressure is different. Yield is minimum when the pressure is kept at low level (50 psi) and the reaction time at high level (15 minutes). Maximum yield is obtained when the pressure and reaction time

Table 7.13 List of process parameters and their levels

Process parameters	Labels	Low level	High level
Temperature	T	80°C	120°C
Pressure	P	50 psi	70 psi
Reaction time	R	5 minutes	15 minutes

Table 7.14 Experimental layout with response values

Run/trial	T	P	R	Yield 1 (%)	Yield 2 (%)	Yield 3 (%)
1	80	50	5	61.43	58.58	57.07
2	120	50	5	75.62	77.57	75.75
3	80	70	5	27.51	34.03	25.07
4	120	70	5	51.37	48.49	54.37
5	80	50	15	24.80	20.69	15.41
6	120	50	15	43.58	44.31	36.99
7	80	70	15	45.20	49.53	50.29
8	120	70	15	70.51	74.00	74.68

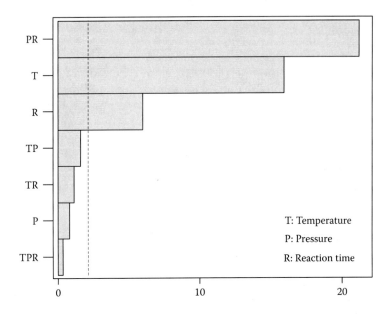

Figure 7.18 Pareto plot of effects for the yield example.

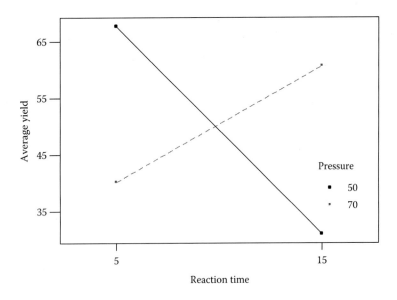

Figure 7.19 Interaction plot between pressure and reaction time.

are kept at low levels. It was quite interesting to observe that the interaction effect was the most critical one for the present example, and this can be easily missed out if one employs or utilises the so-called one-factor-at-a-time or OFAT approach to experimentation. The interaction effects between temperature and pressure as well as temperature and reaction were not found to be significant.

7.23   Process mapping

7.23.1   What is a process map?

Process mapping is the visual representation of a process or several combined processes. Process maps enable improvement of the processes under study by identifying opportunities for improvement (Ward 2007). Though the study begins with a wider viewpoint of the entire process significance, it is later narrowed down to individual steps as the study progresses. Process mapping involves steps such as understanding, documentation, analysis and improvement of processes (Anjard 1996).

7.23.2   When do we use a process map?

The sequence of a process under study is identified using process mapping. Process maps may be attributed to any product or service (Okrent and Vorkurka 2004). They are used to evaluate the performance of individuals or teams and to validate work procedures. The purpose of process mapping is to visually represent the present state of a process and identify improvement opportunities to achieve improved product or service quality, as well as customer satisfaction. Process mapping helps in identifying appropriate practices and benchmarking the processes for ensuring better product sales.

Process mapping can be used to

- Show unexpected complexity, problem areas, redundancy, unnecessary loops and where simplification and standardisation may be possible
- Compare and contrast the actual vs the ideal flow of a process to identify improvement opportunities
- Allow a team to come to agreement on the steps of the process and to examine which activities may impact the process performance

7.23.3   How do we construct a process map?

Step 1: Decide the scope of the process; in other words, where does the process start and where does it end?

Step 2: Ask participants to record individually from their own perspective each step in the process, concentrating on what happens 'most of the time'.

Step 3: Participants should then stick the post-it notes onto the wall; duplicate steps should be placed under one another.

Step 4: As participants work through this process, issues are generated and should be captured on a separate flipchart, as they will prove very valuable later.

Step 5: Use SYMBOLS to describe (map) the process:

Oval symbol denotes start and end of the process

Rectangle symbol represents tasks or activities involved in the process

Arrow symbol denotes direction of process flow

Diamond symbol denotes decision during the process

Step 6: Once people share a common understanding of the problems with a current process, they generate jointly agreed solutions, which can be captured on a different flipchart. A thorough brainstorming session is strongly recommended here for improvement of the current process.

7.23.4   *An illustrative example*

Figure 7.20 shows a simple example of a process map for a pump manufacturing process.

7.24   *Measurement System Analysis*

7.24.1   *What is MSA?*

Measurement and data collection are an integral part of any LSS project. The collected data from any process exhibit certain variability. Some of the observed variability is the inherent variability in the process, and the remaining variability is due to the measurement system variability. A measurement system consists of three components, viz, instruments or gages, inspectors and parts or components under study. Hence, measurement system variability is contributed by these three factors and their interactions. A comprehensive study of the measurement system variability to assess the adequacy of a measurement system is known as measurement system analysis (MSA).

There are two components of uncertainty in MSA: repeatability and reproducibility. This combination is labelled as the precision of a measurement system. Repeatability is the variation found in a measurement system when the same item is measured over and over again without

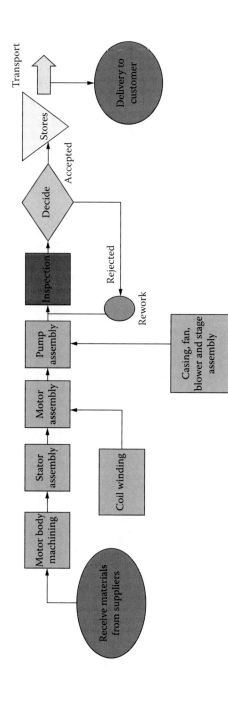

Figure 7.20 Example of a process map. *Note:* A process map is a visual representation of the process. It depicts the sequence of the process from beginning to end. It is used to diagnose any problems existing in the process. A value stream map, on the other hand, provides understanding of how the product flows from the receipt of a customer order until the product is delivered to the customer. It is used to identify the steps that add value and do not add value to the customer. It provides visualisation of both information and product flow.

changing its position or who appraised it, and all at the same time. This uncertainty estimate is really the smallest error you can get on a measurement system without fundamentally changing the equipment or the measurement process. Reproducibility is the variation found when trying to reproduce the measurement under different conditions. These different conditions will include the difference in appraisers, the difference in positioning of the item in the measurement tool, different times and different calibrations. When this value is high, it implies that the measurement process is inadequate (https://www.moresteam.com/toolbox/measurement-system-analysis.cfm).

7.24.2   When do you use measurement system analysis?

MSA is performed before the collection of data from any process. This analysis ensures that the variation due to the measurement system is assessed and controlled before the measurements are taken. Thus, MSA plays a vital role in segregating the inherent process variation and measurement system variation. Hence, it is advisable to use the MSA before the data collection during the measure phase of the LSS project.

7.24.3   How do we create measurement system analysis?

During the MSA study, one tries to estimate the repeatability (difference in readings when measurements are repeated under same condition) and reproducibility (difference in readings when measured by a different operator) of the measurement system, generally known as gage repeatability and reproducibility (gage R&R). This is represented as a percentage of the total variability (AIAG 2002).

The steps for conducting MSA to estimate gage R&R are as follows:

- Select the gage for conducting the study and identify at least two operators for measurement.
- Select 10 parts or units to be measured.
- Each operator measures each part or unit at least two times (equal number of times), and data are recorded. Thus, if there are 10 parts, measured by two inspectors twice, there will be a total of 40 observations.
- To perform the analysis, part numbers, operators and measured data are entered in three different columns of Minitab software.
- Use the Minitab command of 'Stat > Quality Tools > Gage Study > Gage R&R Study (Crossed)' to analyse the MSA data.
- Look at the output of the Minitab analysis and take the appropriate decision. The criteria for acceptance of a measurement system are as follows:

- If gage R&R value is less than 10% of total process variation, the measurement system is acceptable.
- If gage R&R value is between 10% and 30% of total process variation, the measurement system is acceptable depending on the application, cost of the measurement device, cost of repair, etc.
- If gage R&R value is more than 30% of total process variation, the measurement system is unacceptable and should be improved.

7.24.4   An illustrative example

In this example, we consider two operators A and B who have been using a micrometer to measure the thickness of 10 parts. Each part is measured twice, and the data are given in Table 7.15.

The data recorded in Table 7.15 are analysed by the Minitab software, and the results are given in Table 7.16.

From Table 7.16, the total gage R&R is 55.04%. Since this value is beyond the acceptable limit of 30%, the measurement system is not adequate for further collection of data. Actions like calibrating instruments, training inspectors or validating the procedure for inspection must be considered to improve the measurement system. After the improvement actions, it is recommended to repeat the MSA study to reconfirm the improvement in the measurement system.

Table 7.15 Data for gage R&R study

Operator	Part no.	Part number									
		1	2	3	4	5	6	7	8	9	10
A	1	0.09	0.10	0.11	0.08	0.09	0.08	0.08	0.08	0.09	0.08
	2	0.10	0.11	0.10	0.09	0.08	0.08	0.08	0.09	0.08	0.09
B	1	0.09	0.10	0.10	0.08	0.09	0.08	0.09	0.09	0.08	0.08
	2	0.10	0.10	0.10	0.07	0.09	0.08	0.09	0.09	0.08	0.08

Table 7.16 Output of MSA in Minitab

Source	Standard deviation	% Study variation
Total gage R&R	**0.0053820**	**55.04**
Repeatability	0.0053820	55.04
Reproducibility	0.0000000	0.00
Part-to-part	0.0081638	83.49
Total variation	0.0097782	100.00

If the measurement system has error in excess of 30%, the first step to improve results is to analyse the breakdown of the error source. If the largest contributor to error is repeatability, then the equipment or gage must be improved. Likewise, if reproducibility is the largest source of error, appraiser training and adherence to standard procedures can yield improvement.

7.25　Solution Selection Matrix

7.25.1　What is an SSM?

A solution selection matrix (SSM) is a matrix that helps to identify the best solution(s) among several solutions identified, by weighting the impact of each solution on a set of criteria, hence measuring the effectiveness of solving the problem. It helps a team of people to narrow down options through a systematic approach of comparing solutions for a problem by selecting, weighting and applying criteria. For quality improvement activities, an SSM can be useful in selecting a project and in evaluating which solutions or decisions are the most viable.

7.25.2　When do you use an SSM?

An SSM is useful

- When a list of options must be narrowed to one choice
- When the decision must be made on the basis of several criteria
- When you have a number of good alternatives to choose from, and many different factors to take into account

7.25.3　How do you use an SSM?

An SSM works by getting you to list your all possible solutions as rows on a table, and the criteria you need consider as columns. You then score each solution combination, weight this score by the relative importance of the criteria, and add these scores up to give an overall score for each solution:

Step 1: Brainstorm the evaluation criteria appropriate to the solution. If possible, involve customers in this process.
Step 2: Discuss and refine the list of criteria. Identify any criteria that must be included and any that must not be included. Reduce the list of criteria to those that the team believes are most important. Tools such as multi-voting may be useful here.

Step 3: List all of your possible solutions as the row labels on the table, and list the criteria that you need to consider as the column headings.

Step 4: Score each solution from 0 (poor) to 5 (very good) against the criteria. Note that you do not have to have a different score for each solution – if none of them is good for a particular criterion in your decision, then all options should score 0.

Step 5: The next step is to work out the relative importance of the criteria in your decision. Show these as numbers from, say, 1 to 9, where 1 means that the criterion has low importance in the final decision, 5 means moderate importance for the criterion and 9 means high importance for the criterion. It is perfectly acceptable to have criteria with the same importance. The team can assign any score between 1 and 9 depending on the importance of a particular criterion.

Step 6: Now multiply each of your scores from step 4 by the values for relative importance of the criteria that you calculated in step 5. This will give you weighted scores for each option combination.

Step 7: Finally, add up these weighted scores for each of your options. The option with the highest score will not necessarily be the one to choose, but the relative scores can generate meaningful discussion and lead the team towards consensus.

7.25.4   An illustrative example

An LSS black belt has come up with four potential solutions to a problem related to one of the core processes in a manufacturing company which produces automotive parts and accessories for a large car manufacturer. It was not very straightforward for the LSS black belt to choose the right solution as there was a trade-off across a number of criteria. The black belt has decided to utilise an SSM to help him with the decision-making process in this example. The following criteria were initially chosen, which could potentially influence the solution:

- Ease of implementation
- Impact on customer satisfaction
- Impact on business performance
- Potential risks
- Cost of implementation
- Cost to maintain
- Time until solution is fully implemented
- Return on investment
- Enthusiasm of team members
- Potential effects of the solution on other processes/systems

Table 7.17 Solution selection matrix

Criteria	I (6)	S (9)	C (7)	R (6)	Total score
Potential solutions					
Solution 1	12 (6×2)	27 (9×3)	35 (7×5)	18 (6×3)	92
Solution 2	24 (6×4)	18 (9×2)	28 (7×4)	18 (6×3)	88
Solution 3	12 (6×2)	36 (9×4)	14 (7×2)	24 (6×4)	86
Solution 4	24 (6×5)	45 (9×5)	21 (7×3)	24 (6×4)	114

Note: Each solution was mapped against the criteria, and a score was assigned. For instance, solution 4 was easy to implement, and hence a score of 5 was assigned against ease of implementation. Solution 4 had a big impact on customer satisfaction, and therefore a score of 5 was assigned against impact on customer satisfaction. However, solution 4 was not cheap to adopt, and a score of 3 was assigned against that criteria. The risk associated with solution 4 was low, and the team had decided to assign a score of 4 to this. The solution with the highest total score was considered to be the best solution, and according to Table 7.17, solution 4 was the best choice.

A multi-voting analysis was performed on the above criteria, and it was decided to choose just four criteria for choosing the solution. The four criteria chosen were

1. Ease of implementation (I)
2. Impact on customer satisfaction (S)
3. Cost of implementation (C)
4. Potential risks (R)

Table 7.17 presents all the possible solutions and the criteria with a weightings range from 1 to 9 (1 = low importance, 5 = medium importance and 9 = high importance). Also, score each solution from 0 (poor) to 5 (very good).

7.26 Summary

This chapter provides a list of LSS tools which can be used within the LSS methodology for problem solving. The application of tools ranges from very basic ones such as process mapping, CTQs and CTQ tree, all the way through to very advanced techniques such as DoE and FMEA. We believe that each tool covered in this chapter has a specific role to play in a problem-solving scenario, and success depends on the application of the right tool at the right time with the right attitude. The power of the LSS methodology lies with the integration of tools and techniques in a sequential and disciplined manner across the DMAIC methodology. The author would like to conclude this chapter with a simple table (see Table 7.18) showing what tools are best to use in each phase of the methodology.

Table 7.18 Summary of tools and techniques within the DMAIC methodology

Define	Measure	Analyse	Improve	Control
Project charter	VSM	Histogram	Design of experiments	Standard operating
SIPOC	Process mapping	Cause and effect analysis	Failure mode and effect	procedures
VSM	CTQs and CTQ tree	Root cause analysis	analysis	Visual management
VOC analysis		Pareto analysis	Kanban system	5S
		Scatter plot	Solution selection	Control charts
		Correlation analysis	matrix	Run charts
		Run charts	SMED	Poka-Yoke
		Control charts		
		Hypothesis tests		
		Regression analysis		
		Failure mode and effect analysis		

References

Adams, A., Kiemele, M., Pollack, L. and Quan, T. (2004). *Lean Six Sigma: A Tools Guide.* 2nd ed. Colorado Springs, CO: Air Academy Associates.

AIAG (2002). *Measurement Systems Analysis, Reference Manual.* 3rd ed. Southfield, MI: Automotive Industry Action Group.

Anjard, R. P. (1996). Process mapping: One of three, new, special quality tools for management, quality and all other professionals. *Microelectronics Reliability* 36(2): 223–225.

Asaka, T. and Ozeki, K. (1990). *Handbook of Quality Tools: The Japanese Approach.* Cambridge, MA: Productivity Press.

Doolen, T. L. and Hacker, M. E. (2005). A review of lean assessment in organizations: An exploratory study of lean practices by electronics manufacturers. *Journal of Manufacturing Systems* 24(1): 55–67.

Draper, N. R. and Smith, H. (2003). *Applied Regression Analysis.* 3rd ed. New York: John Wiley.

Dvorak, P. (1998). Poka-Yoke designs make assemblies mistake-proof. *Machine Design* 70(4): 181–184.

Evans, J. R. and Lindsay, W. M. (2005). *An Introduction to Six Sigma and Process Improvement.* Mason, OH: Thomson South-Western.

Gopalakrishnan, N. (2010). *Simplified Lean Manufacture: Elements, Rules, Tools and Implementation.* New Delhi: PHI Learning.

Gopalakrishnan, N. (2012). *Simplified Six Sigma: Methodology, Tools and Implementation.* Chennai: PHI Learning.

Haslett, T. and Osborne, C. (2000). Local rules: Their application in a kanban system. *International Journal of Operations and Production Management* 20(9): 1078–1092.

Hines, P. and Rich, N. (1997). The seven value stream mapping tools. *International Journal of Operations and Production Management* 17(1): 46–64.

Kiemele, M. J., Schmidt, S. R. and Berdine, R. J. (2000). *Basic Statistics: Tools for Continuous Improvement*. 4th ed. Colorado Springs, CO: Air Academy Press and Associates.

Lockton, D., Harrison, D. and Stanton, N. A. (2010). The design with intent method: A design tool for influencing user behavior. *Applied Ergonomics* 41(3): 382–392.

Mahapatra, S. S. and Mohanty, S. R. (2007). Lean manufacturing in continuous process industry: An empirical study. *Journal of Scientific and Industrial Research* 66(1): 19.

Montgomery, D. C. (1991). *Design and Analysis of Experiments*, 3rd ed. New York: John Wiley.

Montgomery D. C. (2009). *Introduction to Statistical Quality Control*. 6th ed. New York: John Wiley.

Montgomery, D. C. and Peck, E. A. (1982). *Introduction to Linear Regression Analysis*. New York: John Wiley.

Montgomery, D. C. and Runger, G. C. (2007). *Applied Statistics and Probability for Engineers*. 4th ed. New York: John Wiley.

Moreira, A. C. and Pais, G. C. S. (2011). Single minute exchange of die: A case study implementation. *Journal of Technology Management and Innovation* 6(1): 129–146.

Murata, K. and Katayama, H. (2010). Development of Kaizen case-base for effective technology transfer: A case of visual management technology. *International Journal of Production Research* 48(16): 4901–4917.

Okrent, M. D. and Vokurka, R. J. (2004). Process mapping in successful ERP implementations. *Industrial Management and Data Systems* 104(8): 637–643.

Pyzdek, T. and Keller, P. A. (2003). *The Six Sigma Handbook*, Vol. 486. New York: McGraw-Hill.

Rojasra, P. M. and Qureshi, M. N. (2013). Performance improvement through 5S in small scale industry: A case study. *International Journal of Modern Engineering Research* 3(3): 1654–1660.

Shingo, S. (1986). *Zero Quality Control: Source Inspection and the Poka-Yoke System*. Cambridge, MA: Productivity Press.

Stewart, D. M. and Melnyk, S. A. (2000). Effective process improvement: Developing Poka-Yoke processes. *Production and Inventory Management Journal* 41(4): 48–55.

Tapping, D., Luyster, T. and Shuker, T. (2002). *Value Stream Management: Eight Steps to Planning, Mapping, and Sustaining Lean Improvements*. New York: Productivity Press.

Ward, A. C. (2007). *Lean Product and Process Development*. Cambridge, MA: Lean Enterprise Institute.

chapter eight

LSS project selection

8.1 What is an LSS project?

Lean Six Sigma (LSS) is a methodology that provides insight into processes and enables us to improve these processes. The processes requiring improvement are to be selected based on issues related to increasing market penetration or improving the organisational speed of doing things or reducing the cost of running the business. Hence, the primary goal of LSS is to improve customer satisfaction, and thereby profitability, by improving yield through elimination of defects. Thus, LSS drives for variation reduction and process improvement based on data and their analysis. This LSS methodology is implemented in organisations through a project-by-project approach. First, the area requiring improvement is identified, and then it is framed as an LSS project. This project is then allotted to a team to execute it. The team works on this project and completes it within the time frame.

Since each LSS project requires a huge investment in terms of time and resources, it needs to be ensured that projects requiring such levels of detailed study only are selected as LSS projects. Hence, the problems where solutions are obvious should not be considered as LSS projects. In other words, when the root cause of a problem is unknown from the start of problem solving, this would be a good candidate for an LSS project (Breyfogle 2003).

8.2 Project selection and prioritisation

The green belts (GBs) and yellow belts (YBs) are the backbone of any LSS implementation in small and medium sized enterprises (SMEs). Hence, the organisation should ensure that GBs/YBs pursue the project with a lot of responsibility and passion. This can be achieved only if top management ensures that a suitable environment is created within the organisation which is conducive for result-oriented teamwork. For this, the management should provide the right support and encouragement to the GBs and YBs from the project selection stage to completion of LSS projects.

There can be quite a few problems in every organisation. During the project selection stage, the question that needs to be answered is which problem or issue should be selected as an LSS project. In a few organisations, people generally believe that LSS projects are selected only in those cases where huge financial savings are possible. In this process, quite a few critical problems in the organisations are not considered under LSS methodology. Hence, it should be debated within management about what type of projects should be selected as LSS projects.

Before the selection of projects, it is a good idea to work out a strategy for LSS project selection. This should address two important questions, viz, who will select the project and how the projects will be selected. Instead of individuals selecting a project, it is better to form a team for the finalisation of projects. This team can consist of members of top management and project champions/sponsors. The process owners of the respective processes also can be included in the team. The presence of top management is essential as they understand the priority of the organisation better than anyone else. It is good to include a representative of the finance/accounts department in this team. Their involvement will be helpful even in the estimation of financial benefits due to the LSS projects, as they understand the language of money better (Kumar et al. 2009).

When the projects are selected, generally two approaches can be followed. Either the top management team can select a project and give it to the LSS project team for execution, or the LSS project teams can identify a project and approach top management for approval. In either case, the ultimate authority for approval of the project is with top management.

It is not advisable to select any strategic projects of the organisation, as it can take much more time than a usual LSS project in an SME scenario. It is very important to ensure that the first set of projects is operational related and completed on time. Organisations generally have different mechanisms to identify improvement opportunities. Hence, they can adopt a methodology suitable to their working culture and practices. In this chapter, two project selection methods are proposed for SMEs. The first method is based on effort–impact matrix and the second one is based on criteria-based prioritisation.

Prioritising projects based on effort vs impact matrix is the simplest method for selecting projects. In this approach, one tries to estimate the effort in completing a project and its impact (benefit) for the organisation. The project duration and resources including manpower and other budgetary support required for successful completion of the LSS project are considered as 'effort'. The effect of results on the processes and business is estimated as the 'impact'. Hence, the impact can be cost reduction, on-time delivery, return on investment, etc. Generally an effort–impact matrix is prepared after classifying the effort and impact in two categories viz, 'low' and 'high'. A typical form of effort–impact matrix is presented in Figure 8.1.

Now, from Figure 8.1, we have four combinations of projects based on 'high' and 'low' values of effort and impact. The first set of LSS projects can be selected from the category of 'low effort' and 'high impact'. During

Figure 8.1 Effort–impact matrix.

the initial period of implementation of the LSS initiative, there can be quite a few projects which fall under this category in every organisation. If the teams are selecting such projects, they can achieve faster success. Thus, through these set of projects, the teams can learn the LSS methodology.

The second method for selecting the project is based on various criteria identified from the business. In this approach, the criteria for selecting a project need to be identified first. Then, each of the criteria needs to be given a ranking based on order of preference. Generally, a ranking in a scale of 1 to 10 can be used for this purpose. Thus, the most preferred criterion is given a ranking of 10, and the least preferred criterion is ranked as 1. After a list of possible criteria and ranking for each criterion are finalised, the projects are ranked. All the possible projects are evaluated against the criteria, and the total score is calculated for each project. Finally, the selection of an LSS project is made based on the total score. The highest score represents the most preferred project among the list of projects.

Examples for possible criteria are

- Direct link to strategic goals
- Contribute to bottom-line performance
- Directly benefit key customers and stakeholders
- Can be completed within a 3- to 6-month time frame

An example of criteria-based project selection in an automobile supplier organisation is described below.

First, the team debated about the criteria to be used for project selection and ranked each criterion based on its importance, in a 1 to 10 scale. A total of four criteria were selected in this case. Table 8.1 presents the selected criteria and respective score.

Once the criteria and the score are finalised, the team along with the champion identifies four projects from various processes. Each of these projects are ranked against each criterion, and the values are written in the corresponding cell. Finally, a total weighted score is calculated for each

Table 8.1 Criteria and score

S. no.	Criteria	Score
1	Direct link to strategic goals	9
2	Contribute to bottom-line performance	5
3	Directly benefit key customers and stakeholders	7
4	Can be completed within 3- to 6-month time frame	8

project. For example, the first project 'Reducing inventory of raw material in stores' is ranked against each of the four criteria. Based on importance, a ranking of 7, 6, 2 and 2, respectively, are given by the team against each criterion. Then, these ranks are multiplied by the respective weightage of the criteria. Thus, $7 \times 9 + 6 \times 5 + 2 \times 7 + 2 \times 8$ provides a total score of 123. The details are provided in Table 8.2.

The last column in Table 8.2 provides the total weighted score for each project. Since the second project, namely *'Reducing rework and scrap in grinding operation'*, got the highest total score, it was selected as the LSS project.

Thus, under an SME scenario, one can use the effort–benefit matrix or the criteria-based method for selection of projects. This ensures that important projects are identified from the processes.

8.3 *Management of project reviews*

Review of progress plays a vital role in the success of LSS implementation. Two types of reviews have to be performed for each LSS project. The first type of review is by the top management or a champion, and the second type of review is by the LSS expert. In the context of an SME, an LSS expert is an experienced GB who can guide the LSS projects.

- The focus of review by the top management is about overall progress of the project with respect to time line, project objectives and their alignment with corporate strategy. The smooth functioning of LSS teams is ensured during these reviews. The need for any budgetary or other supports required by the LSS team is also planned at the time of reviews. These reviews can be generally conducted after completing every phase of the define-measure-analyse-improve-control (DMAIC) methodology. The following questions can be included as check points during the review by management or champion:
  - Is the project executed as per the planned schedule?
  - Are team members able to provide the time required for the project?
  - Is the overall progress made in the specific phase of DMAIC acceptable?
  - Is there a problem with respect to budget and resources?

Table 8.2 Criteria and score for project selection

Criteria projects	Direct link to strategic goals (9)	Contribute to bottom-line performance (5)	Directly benefit key customers and stakeholders (7)	Can be completed within 3- to 6-month time frame (8)	Total score
Reducing inventory of raw material in stores	7	6	2	2	123
Reducing rework and scrap in grinding operation	5	9	7	8	203
Improving the first pass yield of injector assembly	5	8	8	3	165
Improving the on-time delivery of injectors to customers	7	3	9	2	157

The review by the LSS expert is mostly based on the methodology of LSS. The adequacy of the methodology, appropriateness of data collection, appropriateness of analysis and interpretation are the main focus during this review. These reviews have to be conducted after every phase. In addition to that, whenever the teams require any guidance in the project, handholding sessions can be planned. The following are a few important points to be checked during this review by the LSS expert:

- Is the project addressing all important steps in different phases of DMAIC methodology (refer to Chapter 6 for details of DMAIC steps)?
- Is the plan for data collection appropriate at every stage?
- Is the method of data collection proper?
- How correctly is the analysis being performed? Are the right tools being used for analysis?
- Are the conclusions from each analysis and the resulting actions appropriate?
- Are the benefits estimated from the project based on the existing procedures of the organization?

The review by management and the LSS expert together will ensure on-time successful completion of the LSS project. Hence, these reviews are very important for any LSS implementation. If there is not enough progress in the project after these reviews, the champion and the LSS expert need to sit down together and discuss what is going wrong in the project. They can try to understand the difficulties/hurdles from the project team and suggest countermeasures to ensure progress in the project.

8.4   Some tips for making your LSS projects successful

The success of the first set of LSS projects is highly critical, as it decides the fate of future implementation of LSS in the organisation. Hence, management should take care to ensure successful and on-time completion of projects (Gijo and Rao 2005). It needs to be noted that, like any other initiative, LSS projects can also fail for many reasons (Gijo 2011). Sustaining the interest and passion of the LSS teams is very important here. The following points should be considered in order to make sure that LSS projects are successful:

- Select projects from critical business/customer issues of the organisation so that everyone in the organisation feels the importance of this problem and provides support for successful completion of the same.

- Involvement of top management in all stages, from project selection to completion of projects, gives a strong signal to the entire organisation about the priority given by management to LSS implementation.
- The scope of the project is suitably defined so that the team will be able to complete it within the time frame provided. If the scope of the project is very large, completing it successfully within the scheduled time is unlikely.
- Selection of appropriate team members plays a very vital role in the success of the LSS project. People who are highly capable of managing change within the organisation, with an eagerness to learn and implement new ideas, are to be included in the team.
- Introduction of appropriate reward and recognition schemes in the organisation can motivate people to come forward and take up more LSS projects. This reward and recognition scheme can include sharing a part of the savings achieved (e.g. 15%) with the team, giving weightage in the annual appraisal system of the company for promotions, etc. It is also suggested that successful GBs can pursue a BB course and train GBs and YBs within the business.
- As LSS places emphasis on data-based decision-making, projects to be selected in processes where data can be collected. If the cycle time of the process is long, then completion of the project within a stipulated time frame is a big challenge.

8.5   Summary

This chapter provides an overview about the LSS projects and the details regarding LSS project selection. The different approaches that can be adopted for selection of LSS projects are discussed. The role of project reviews in successful completion of LSS projects are also discussed in this chapter. The different types of reviews to be performed for LSS projects are presented with a detailed checklist. The chapter is concluded with some tips for making LSS projects successful.

References

Breyfogle, F. W. (2003). *Implementing Six Sigma: Smarter Solutions Using Statistical Methods*. New York: John Wiley.

Gijo, E. V. (2011). Eleven ways to sink your Six Sigma project. *Six Sigma Forum Magazine* 11(1): 27–29.

Gijo, E. V. and Rao, T. S. (2005). Six Sigma implementation: Hurdles and more hurdles. *Total Quality Management and Business Excellence* 16(6): 721–725.

Kumar, M., Antony, J. and Cho, B. R. (2009). Project selection and its impact on the successful deployment of Six Sigma. *Business Process Management Journal* 15(5): 669–686.

chapter nine

Industrial case studies of Lean Six Sigma*

9.1 Case study 1: Application of Six Sigma methodology in a small-scale foundry industry

9.1.1 Background of the company

The company where the case study was performed started its operations around 15 years ago as a small-scale, private limited company for manufacturing automobile leaf springs. This had an initial capacity of 2,000 tons/annum. The capacity has been gradually increased, and as of today, the unit has a capacity of 7,000 tons/annum, with employee strength of 100 people. The sales, which were to the extent of US$100,000 during the first year, have steadily increased to a business turnover of around US$1.5 million. There was no formal quality improvement initiative implemented in the organisation. There were isolated events of Kaizen and Lean methodology implementation in some of its processes in the past. Very small improvements in the processes were achieved through these initiatives.

9.1.2 Background to the problem

The leaf springs are commonly used for suspension in wheeled vehicles and are designed to withstand varying levels of stress and vibration during vehicle movement due to different road conditions (Gijo et al. 2014). Hardness is one of the most important characteristics maintained during manufacturing of leaf springs. The leaf springs with high hardness will make the material brittle, which will result in breakage of the spring; whereas low hardness will not take up specified loads, creating vibrations. Thus, it is very important to manufacture the product within the specified hardness limits. The company follows the quench hardening process for the manufacturing of leaf springs. The company had an increasing problem of rework and rejections in the hardening process, with an approximate rejection of 48.33% during the past 6 months, as the hardness of the manufactured spring was crossing desired specification limits.

---

* The authors have been granted permission to reproduce three case studies published in this chapter from both Emerald and Interscience Publishers.

9.1.3 Six Sigma methodology (DMAIC)

As the company was facing high rejection rates in its processes, the organisation was struggling to manage its commitments to its customers. This situation not only increased the manpower, material and other overhead costs for manufacturing, but it also created the fear of losing business due to lack of on-time delivery of products to the customers. Usage of technical knowledge and Kaizen improvements were not able to identify the root causes of the problem. As the issue was very complex, and the solutions were unknown, the company decided to apply the Six Sigma define-measure-analyse-improve-control (DMAIC) approach in this process.

The remaining part of this section illustrates various activities carried out at different phases of Six Sigma DMAIC methodology.

9.1.3.1 Define phase

After a detailed discussion at various levels within management, a project charter was drawn with all the details of the project as illustrated in Table 9.1. This project charter forms the basis of all future actions and

Table 9.1 Project charter

Project Title: Reducing rejection in the hardening process of automobile leaf spring manufacturing.		
Background and reason for selecting the project: Hardness of the material used for automobile leaf spring manufacturing crosses the required specification of 245–265 BHN, leading to an approximate rejection of 48.33% of the products manufactured for the past 6 months. This increases the material and labour cost in the company and thus affects profitability and on-time delivery.		
Aim of the project: To reduce the rejection of the hardening process from 48.33% to less than 5%.		
Project Champion	General Manager	
Project Leader	Black Belt	
Team Members	Engineer–Production, Engineer–Quality Control, Manager–Production, Supervisor–Production, Operator–Shift I, Operator–Shift II	

Characteristics of product/process output and its measure

CTQ	Measure & Specification	Defect Definition
Hardness	245–265 BHN	Hardness Crossing 245–265 BHN
Expected Benefits	Reduction in rejection and rework as a result of reduced variation in hardness. This will help the organisation to improve the on-time delivery of products to its customers.	
Schedule	Define: 1 week, Measure: 1 week, Analyse: 2 weeks, Improve: 2 weeks, Control: 4 weeks	

Table 9.2 SIPOC

Supplier	Input	Process	Output	Customer
Shearing department	Material	Hardening process	Hardened leaf spring	Hardness inspector
Heating shop	Furnaces		Production report	General manager
Quenching shop	Reservoir			
	Quenching oil			

Hardening process

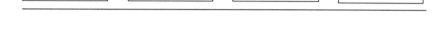

decisions within the project. Next, a supplier-input-process-output-customer (SIPOC) analysis was performed (Table 9.2).

The need of the project is to reduce the rejection percentage of the process, which is at 48.33% for the past 6-month period. Whenever the hardness values fall beyond the specification limits of 245–265 BHN, the product is rejected. Hence, if the variation in hardness is reduced, the rejections and rework could be eliminated. Thus, the team decided to focus on hardness for further improvement, which is defined as the critical to quality (CTQ) characteristic for the project. The specification limit for this hardness measurement was 245–265 BHN.

9.1.3.2   Measure phase

The first step in the measure phase is to evaluate the adequacy of the measurement system used for the collection of data. Hence, the team decided to conduct a measurement system analysis (MSA) for the measurement system used in recording the data on hardness. The instrument used for measuring hardness is a 'Brinell hardness testing machine' with least-count of 0.001 mm. The total gage repeatability and reproducibility (total gage R&R) value was found to be 7.59%, which is within the acceptable limit of 10%.

Next, a detailed data collection plan was prepared with sample size and type of sampling with stratification factors like operator, shift, etc. As per the plan, the data were collected based on hardness generated from the process. The data were tested for normality with the 'Anderson–Darling normality test', and the p-value of the test was found to be less than 0.05, which led to the conclusion that the data were from a population that is not normal. Since none of the distributions were fitting to this data, and transformations like Box–Cox transformation and Johnson transformation failed to transform these data to normality, the 'observed PPM total' of 270,000 from Figure 9.1 was considered as an estimate of the baseline performance of the process.

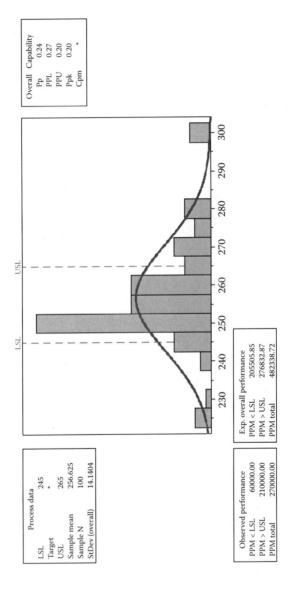

Figure 9.1 Process capability analysis of hardness.

9.1.3.3   Analyse phase

A brainstorming session was planned and conducted by the team with the involvement of all the concerned personnel in the process, and a list of potential causes for variation in CTQ was generated. A cause and effect diagram was drawn based on these causes and is presented in Figure 9.2.

All the causes listed in the cause and effect diagram in Figure 9.2 are to be validated based on data to identify the root causes. Hence, it is necessary to explore the type of data that is possible to collect on each of these causes and plan for an appropriate analysis to make meaningful conclusions about the potential causes. A plan was prepared with details on each cause and the type of validation required. A summary of these details is presented in Table 9.3. As per the plan given in Table 9.3, a few of the causes are to be validated by process observation or 'Gemba', and the remaining causes are to be validated based on various statistical analyses. There are a few cases where Gemba analysis has been used to identify the root causes, and those will be explained in the following paragraph. A summary of the results of all these validations is also presented in Table 9.3. Process parameters like reservoir volumes, quenching time, oil temperature, type of quenching oil, etc., were fixed during the establishment of the process based on the trial-and-error method. Hence, it was decided to conduct a design of experiment (DoE) at the improvement phase for these parameters so that an optimum process setting could be identified.

Some of the Gemba observations during this study were as follows. Profiles of the operators and time and motion study for a period of 1 week revealed that the operators have adequate skills and experience to carry out the process. It was also observed that the workload and rest times

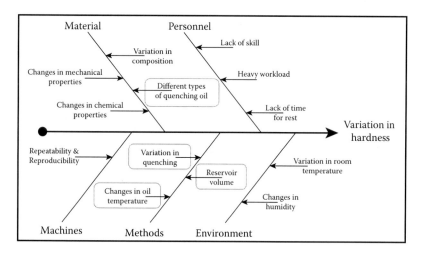

Figure 9.2 Cause and effect diagram.

Table 9.3 Cause validation details

S. no.	Causes	Specification/desired status	Observation	Remarks
1	Reservoir volume	Small ($1.5 \times 2 \times 1$) m³–medium ($1.75 \times 2.25 \times 1$) m³–large ($2 \times 2.5 \times 1$) m³	Medium reservoir was used where space not sufficient	To be studied by DoE
2	Variation in quenching time	(4–6) min	Variation beyond specification limits was observed	To be studied by DoE
3	Changes in oil temperature	(35–45)°C	Variation beyond specification limits was observed	To be studied by DoE
4	Lack of skill	Operator must be at least semiskilled	Skill of the operator is adequate	Not a root cause
5	Heavy workload	No more than 48 hours duty/week, with weekly holiday	Workload is distributed as per plan	Not a root cause
6	Lack of time for rest	Half an hour rest time for every 5 hours	Rest times are allotted as per industrial norms	Not a root cause
7	Repeatability & reproducibility	% variation in gage R&R <30%	Within the acceptable limit	Not a root cause
8	Variation in the room temperature	Should be less than 35°C	Except the morning time, it was observed as constant	Not a root cause
9	Changes in humidity	40%–60% RH	No significant variation	Not a root cause
10	Variation in composition	Should be of same composition	Constant, as it was checked before purchase	Not a root cause
11	Different types of quenching oil	No specification available	Slow cooling rate	To be studied by DoE
12	Changes in the mechanical properties	Martensite structure	No variation found	Not a root cause
13	Changes in the chemical properties	Properties related to Martensite structure	No variation found	Not a root cause

are strictly followed as per regulatory requirements. Repeatability and reproducibility of the measurement system was confirmed with an MSA study during the measure phase. Room temperature and humidity were checked every hour for 1 week, and it was confirmed that both are within the desired specification limits. Material composition was confirmed from the past months' raw material purchase records, as one component in every batch was checked by the purchase officer before lot acceptance. Mechanical and chemical properties are verified with microstructure study by taking two components/day for a period of 1 week. The conclusions from these analyses are also included in Table 9.3.

9.1.3.4   Improve phase

As per the decision of the team in the analyse phase, a DoE was planned and conducted during this phase to identify the optimum settings for the process parameters. After a detailed discussion, the parameters selected for experimentation were quenching time, oil temperature, type of quenching oil and reservoir volume. During the brainstorming session, the team felt that the interaction of 'quenching time' with 'reservoir volume', 'oil temperature' and 'type of quenching oil' could have a significant impact on hardness. Hence, these interactions also were considered for further study. The response of the experiment was decided as hardness measured on the component with the hardness tester. The selected factors and their levels for experimentation are presented in Table 9.4. The design layout for the experiment was prepared by allocating the factors and level to the $L_{27}(3^{13})$ orthogonal array (OA). The experimental sequence given in the design layout was randomised, and experimentation was completed while hardness values were recorded. The experimental plan and the collected data are presented in Table 9.5.

Since the variation in the output characteristic (hardness) was studied and reduced for this process, Taguchi's signal-to-noise (S/N) ratio concept was utilised for analysing the data. Since hardness is a nominal-the-best type of characteristic, the S/N ratio formula used for analysis

Table 9.4 Factors and their levels for experimentation

		Level		
S. no.	Factor	1	2	3
1	Quenching time (in min)	4	5	*6
2	Oil temperature (in °C)	35	40	*45
3	Quenching oil (quench-o-meter rating, in sec)	*Fast (9)	Medium (13)	Slow (17)
4	Reservoir volume (in m³)	Small (1.5×2×1)	*Medium (1.75×2.25×1)	Large (2×2.5×1)

* Existing level.

Table 9.5 Experimental plan with data

Exp. no.	Quenching time	Oil temperature	Quenching oil	Reservoir volume	Hardness		
					1	2	3
1	4	35	Fast	Small	290	294	300
2	4	35	Medium	Medium	270	272	270
3	4	35	Slow	Large	277	277	276
4	4	40	Fast	Medium	275	270	275
5	4	40	Medium	Large	270	270	271
6	4	40	Slow	Small	270	270	268
7	4	45	Fast	Large	263	265	262
8	4	45	Medium	Small	258	258	257
9	4	45	Slow	Medium	262	260	262
10	5	35	Fast	Small	260	259	261
11	5	35	Medium	Medium	255	256	255
12	5	35	Slow	Large	257	256	257
13	5	40	Fast	Medium	255	255	256
14	5	40	Medium	Large	248	249	248
15	5	40	Slow	Small	250	251	252
16	5	45	Fast	Large	243	244	245
17	5	45	Medium	Small	238	230	239
18	5	45	Slow	Medium	242	243	240
19	6	35	Fast	Small	242	230	239
20	6	35	Medium	Medium	239	230	237
21	6	35	Slow	Large	238	220	237
22	6	40	Fast	Medium	237	240	236
23	6	40	Medium	Large	230	231	232
24	6	40	Slow	Small	232	220	230
25	6	45	Fast	Large	225	226	227
26	6	45	Medium	Small	220	230	220
27	6	45	Slow	Medium	220	224	220

was $10\log(\bar{Y}^2/s^2)$, where $\bar{Y}$ is the average and s the standard deviation for each experiment. The S/N ratio values were calculated for all the 27 experiments, and these S/N ratio values were subjected to an analysis of variance (ANOVA) to identify the significant factors and interactions. The ANOVA table thus obtained is presented in Table 9.6. From the ANOVA table, it was found that the p-values for factors 'quenching time' and 'reservoir volume' and the interaction of 'quenching time' with 'quenching oil' and 'oil temperature' were found to be less than 0.05, leading to the conclusion that these factors and interactions significantly affect the hardness. The optimum condition was identified from the main effect and interaction plots for the S/N ratio values and is presented in Table 9.7.

Table 9.6 ANOVA table

Source	DF	SS	MS	F	p-value
Quenching time	2	699.91	349.96	23.24	0.001*
Oil temperature	2	56.94	28.47	1.89	0.231
Quenching oil	2	20.12	10.06	0.67	0.547
Reservoir volume	2	249.96	124.98	8.30	0.019*
Quenching time * Oil temperature	4	289.03	72.26	4.80	0.044*
Quenching time * Quenching oil	4	304.23	76.06	5.05	0.040*
Quenching time * Reservoir volume	4	54.84	13.71	0.91	0.514
Error	6	90.37	15.06		
Total	26	1765.40			

* Significant at 5% level of significance.

Table 9.7 Optimum factor level combination from the experiment

Sl. No	Parameters	Optimum level after DoE
1	Reservoir volume	Large ($2 \times 2.5 \times 1$) m^3
2	Quenching time	5 min
3	Oil temperature	40°C
4	Quenching oil	Medium (13 sec)

Finally these optimum results were implemented after preparing an implementation plan with responsibility and target dates. The results were observed after successful implementation of the solutions for a period of 1 week. The data on hardness were recorded during implementation. These data were analysed to determine the level of improvement in the process characteristics. The PPM level of hardness was found to be 7924. The overall rejection percentage was reduced from 48.33 to 0.79, which was very significant for the process. The individual control chart (Figure 9.3), plotted for comparing the process data before and after the project, shows significant reduction in variation in hardness after the project.

9.1.3.5   Control phase

In order to ensure that the proposed methods of improvement are sustained, the team implemented a set of control mechanisms. The process is standardised and is documented in quality management system (QMS) documents. Also, a process flow chart was prepared and displayed at the shop floor with details of process specifications. This display helps everyone to understand the process details. Check sheets were prepared for data collection, and control charts were made to

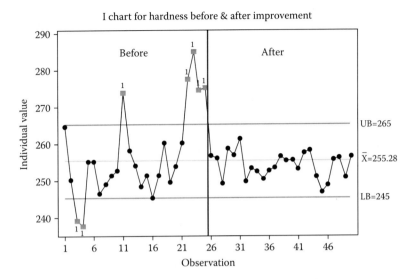

Figure 9.3 Individual chart for hardness.

monitor the process so that the operator can take timely action before the critical process parameters and performance characteristics go out of specified limits.

A periodic review of these results was planned to ensure sustainability of the achieved results. An 'individual–control chart' for hardness was introduced for monitoring the process, along with an out of control action plan (OCAP). This OCAP helps the operator to initiate action in the process in case of assignable causes. It is also necessary to make sure that all the employees are aware of the improvement actions implemented in the process. Hence, a 1-day awareness training program was arranged for all the employees about the Six Sigma methodology.

9.1.4 Managerial implications

This case study was an eye opener for the management as it displayed a significant improvement in the process. Data and their analysis gave confidence to the people and the top management for taking decisions about the process. This has changed the mindset that 'it is not invented here, hence not applicable to our process'. The success in this project has made them the 'change agents' in the process of cultural transformation of the organisation. There were isolated efforts in the organisation in the past to implement initiatives like statistical process control, quality circles, small group activities, Kaizen, etc. During the implementation of those initiatives, no systematic effort was made to identify the improvement opportunities in line with business priorities or customer requirements. As a result, the impact was not very visible in the organisation, whereas in Six Sigma, projects were identified with respect to the voice of the business and the customer, and the problems addressed were of highest priority to the organisation. Hence, management decided to use Six Sigma methodology for all future improvement initiatives in the organisation.

For monitoring of Six Sigma initiatives, a core group was formed with all functional heads of the organisation. The responsibility of this team was selection of projects and monitoring the execution of projects. All issues related to implementation were also reported to this team for further action. Thus, Six Sigma was introduced as a system in the organisation to address any type of problems in the processes. The ultimate objective of management was to bring a cultural change in the organisation by involving everyone in the organisation in its movement towards excellence.

9.1.5   Key lessons learned from the case study

The learnings from this initiative are summarised for implementing future improvement activities effectively. Key lessons learned from the case study focus around the leadership activities, involvement of people in improvement initiatives, data collection and subsequent data-based cause validation. During define, analyse and improve phases of the study it was observed that all hurdles in executing the study were cleared by strong leadership at middle management level. Thus, it was identified that improvement initiatives require strong leadership support not only at higher levels but also within middle levels of the organisation.

One of the reasons for the success of this study was the strong support from the champion. The champion was keen to implement the Six Sigma methodology for addressing process problems in the organisation. Training imparted during the measure phase on technical details of the process under study, data collection plans and some of the key tools of the Six Sigma approach were the backbones in achieving the project goal. Involving the people from all levels in the organisation in the development of the data collection process and sharing the inferences from data analysis with them greatly helped in getting support for collecting the data. Quite often the shop floor workers are not aware about the technical details of the process they are working with. Proper training on the technical details will help them to do process based thinking, recognise variation in the critical characteristics and focus on breakthrough improvement in performance. During the improve phase, support from all levels of the organisation is required for successful implementation of the solutions.

9.1.6   Recap of tools used

The tools used during this case study:

Analysis of variance
Anderson–Darling normality test
Cause and effect diagram
Control chart
Design of experiments
Gage repeatability and reproducibility study
Gemba analysis
Main effect plot
Orthogonal array
Process capability evaluation
Signal-to-noise ratio
SIPOC
Taguchi method

9.1.7   Summary

The case study reported in this section is the application of Six Sigma DMAIC methodology in improving the leaf spring manufacturing process of a foundry shop. The root causes for the problem of rejection and rework were identified through data-based analysis at different stages in the case study. The process parameters were optimised, and measures for sustainability of the results were incorporated in the process. As a result of this study, the overall rejection was reduced from 48.33% to 0.79%, which was a remarkable achievement for this small-scale industry. The company has invested around US$1100 for the project towards training and other activities. This in turn resulted in a financial savings of US$8000/year due to the reduction of rejections in the process. This substantial benefit resulted in reduced material scrap rate and decreased lead time of the process. This led to the improvement of on-time delivery to the customer. This case study illustrates the step-by-step application of Six Sigma DMAIC methodology in a small-scale foundry industry to solve an age-old problem in the organisation.

9.2 Case study 2: Application of Six Sigma methodology in road construction for wind turbine installation

9.2.1 Background of the company

In today's global economy, the renewable energy sector has a significant role. Wind generated energy has become the fastest growing source of renewable energy. Wind power is expected to grow worldwide in the 21st century. India also has progressed in this direction during the last decade. The company where this case study was executed is a windmill manufacturer in India. This company currently has 275 employees and has been in operation for the last 12 years. This organisation manufactures wind turbines for its customers and manages the installation, commissioning and maintenance of the wind turbines. The turbines are produced at its manufacturing facility and then transported to various locations in the country and installed (Gijo and Sarkar 2013).

9.2.2 Background to the problem

Even though solar energy is a relatively newer industry, the challenges in this area are far beyond the imagination of other businesses. In India, this industry has shown steady growth for the past decade, attracting more players in the market. The cost of doing this business is escalating because of rising land prices and the cost of raw materials. Most windmills are located in geographically fragile locations, which make it even more difficult in terms of the execution of activities. The organisations dealing with wind energy face many challenges because the activities are totally different from the traditional manufacturing set-up and can be described as follows:

1. Identification of location
2. Assessment of wind availability
3. Procurement of land
4. Development of land
5. Installation of wind turbine
6. Servicing of wind turbine for agreed period, generally for 20 years

Customers expect timely commissioning of wind turbines and servicing for an agreed period, which helps them to get their return on investment (ROI). The power generation guarantee (based on wind availability) has to be taken up by the organisation. Any delay on site development, commissioning or servicing affects the power generation/ROI and in turn leads to customer dissatisfaction. One of the major activities which affects the delay is the construction of the road to the project and its maintenance. Through the road, they need to transport the whole turbine in which the blade length is of minimum 26 m. It is a challenge to the organisation in developing the road and maintaining it. The sites are generally on hilltops, in coastal areas, in deserts, etc., all of which add more challenges.

In traditional manufacturing organisations, implementation of any improvement initiative is relatively easy, as the organisation has to plan its implementation during the routine manufacturing operation. However, in the wind energy sector, organisations have to start activities a few years before selling a turbine to a customer. The organisation also has to take care of the maintenance of the equipment for 20 years thereafter. Customers will be satisfied only when all the activities are completed on time, and power generation takes place as agreed upon.

9.2.3   Six Sigma methodology (DMAIC)

In order to improve the overall efficiency and customer satisfaction, Six Sigma methodology is used in the wind energy sector as well. Studies are more challenging in this sector because it is a newer application and because of its large dependency on people and the coordination requirement between various agencies. For installation of wind turbines, the company under study used to construct approximately 500 km of road annually in various windmill project sites across the country. Variation of land properties among the sites was a challenge for making good roads. Also high failure rates (i.e. damage) of the roads make them unusable and inaccessible, which in turn affects the installation and maintenance activities. Maintenance costs for the roads were approximately US\$2.0 million annually because of the high failure rate of the roads. Because of these circumstances, management decided to develop sustainable wind farm roads using Six Sigma methodology. The DMAIC approach was selected for this study, as we are improving an existing process of road construction. The following sections will explain the step-by-step application of the DMAIC approach for completing this study.

9.2.3.1   Define phase

A team was formed which consisted of the project manager as the black belt, a project engineer, two supervisors and a maintenance engineer. The general manager in charge of project execution was identified as champion for the project. During the define phase of the project, the team formed a project charter (Table 9.8) with all details of the project including the project title, problem definition, scope of the project, target benefits and schedule. After defining the project in the project charter format, the team prepared a SIPOC mapping to have better clarity in the process

Table 9.8 Project charter

Project Title: Development of sustainable wind farm roads

Background and reasons for selecting the project:

Wind farm roads are currently wearing out within 3–6 months. Also, there is large variation in the road formation process.

Aim of the project:

To arrive at best practices for making sustainable wind farm roads so that the life of the road is at least 6 months.

Project Champion:	General Manager–Project Execution
Project Leader:	Manager–Projects
Team Members:	Engineer–Projects
	Engineer–Maintenance
	Supervisor–Site Development
	Supervisor–Installation

Characteristics of product/process output and its measure

CTQ	Measure & Specification	Defect Definition
Failure rate of roads:	Life measured in months for each 50 m patch of road and specification is more than 6 months.	Road life less than 6 months.

Expected Benefits:	Reliability in service, reduction in maintenance cost.
Expected customer benefits:	Improvement in customer satisfaction.
Schedule:	Define: 4 weeks, Measure: 6 weeks
	Analyse: 16 weeks, Improve: 8 weeks
	Control: 8 weeks.

---

under consideration for improvement (Table 9.9). This SIPOC provides a high-level view of the process, defining the scope of the Six Sigma project. The objective of the project was to create a system which would result in sustainable construction for the roads leading to wind farms. The team further elaborated this objective by defining 'no-failure within 6 months of construction' of the roads, which will eventually lead to 'reduction in maintenance cost' of the roads. A target of 6 months was selected because during the first 6 months of a wind turbine project, extensive movement of heavy vehicles like cranes, trailers, etc., occurs. Once the project is commissioned, for servicing of wind turbines light vehicle movement is sufficient. The possible failures of the roads include cracks, subsidence, stripping, sub-base failure, slippery surfaces and landslides (Figure 9.4). The team decided to identify CTQs for the project as 'failures within 6 months', and the unit of measure is defined as '50 m patch of road'. A reduction of 10% was targeted for the maintenance cost during this project.

Table 9.9 SIPOC

Suppliers	Inputs	Process	Outputs	Customers
WRD	Micrositing drawing			Installation team
Consultant	Survey drawings			Service team
Land team	Land			
Meteorolgical dept	Meteorological data	Wind farm road construction	Standard and sustainable windfarm road	
Consultant	Soil investigation			
Design team	Construction drawing/ specifications			
Road contractor	Equipment			
Road contractor	Raw materials			

Road Survey	→	Soil investigation	→	Methodology selection	→	Road construction	→	Testing and commissioning

9.2.3.2   Measure phase

The objective of the measure phase in a Six Sigma project is to understand the baseline performance levels of the CTQs selected for the project based on the data collected from the process. Before collecting data from the process, a failure mode and effect analysis (FMEA) was performed to understand the complexity of the process and to decide on further data collection in the process (refer to Table 9.10 for FMEA). Based on the output of FMEA, a data collection plan was prepared with details of characteristics where data need to be collected with stratification factors and sampling methods. The stratification factors identified were failure type and location (different states of India). Pareto charts were plotted with these stratification factors for the collected data (Figure 9.5). The number of units (50 m patch of road) considered for calculation of baseline performance is 4455. Since there were six different types of failures possible, the number of opportunities considered for each unit is six. For these 4455 units of road, the number of defects was identified as 228, resulting in defects per million opportunities (DPMO) of 8530. The corresponding approximate sigma level was found to be 3.9. This was the baseline performance of the process.

9.2.3.3   Analyse phase

The objective of the analyse phase in any Six Sigma project is to identify the root cause(s) of the problem under consideration so that corrective actions can be initiated and improvements can be achieved. Since the team was a cross-functional team for this study, it was first decided to prepare an activity flow chart of the entire process to have a clear understanding of the process by all members of the team. This activity flow chart helped to identify all inefficiencies and bottlenecks of the process. After a detailed study of the process through the flow chart, the team performed a brainstorming session to identify the potential causes of bad roads. The

Figure 9.4 Different types of failures: (a) crack, (b) landslide, (c) stripping, (d) sub-base failure, (e) slippery surface, and (f) subsidence.

potential causes generated through the brainstorming session were presented as a cause and effect diagram in Figure 9.6. These causes were validated through data analysis and Gemba (workplace) investigation. The method of cause validation depends on the relationship between cause and effect. In case the relationship between cause and effect is known and already established, then we need to see the occurrences of causes through workplace or Gemba. Where the relationship is not established

Table 9.10 Potential failure mode and effect analysis

Process Function/ Requirements	Potential Failure Mode	Potential Effect of Failure	SEV	Potential Causes)/ Mechanism(s) of Failure	OCC	Current Process Controls	DET	RPN
Sustainable road construction/No failures within 6 months	Development of various types of defects	Landslide	9	Improper compaction	5	Lab test	5	225
		Water stagnation	5	Absence of toe wall	9	Inspection	5	405
		Increased vehicle operation cost	1	Improper grading	1	Grading guidance	5	45
		Road accident	9	Improper camber	1	Inspection	5	45
				Improper drainage	5	Drainage plan	1	45
				Low relative density of soil	1	No control	10	90
				Improper maintenance	1	Maintenance procedure	1	9
				High stress concentration	9	No control	10	810

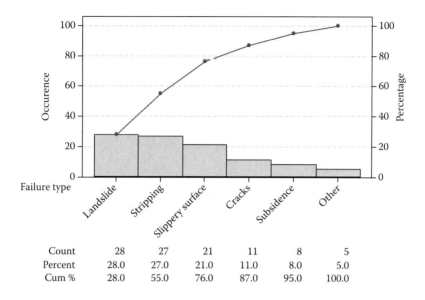

Failure type	Landslide	Stripping	Slippery surface	Cracks	Subsidence	Other
Count	28	27	21	11	8	5
Percent	28.0	27.0	21.0	11.0	8.0	5.0
Cum %	28.0	55.0	76.0	87.0	95.0	100.0

Figure 9.5 Pareto chart for road failure.

or unknown, the cause needs to be validated by collecting data and subsequent analysis through test of hypothesis techniques.

The cause validation in this study was done in two stages. During the first stage, for causes where relationship between cause and effect is unknown, hypothesis tests were performed to validate the cause. For the remaining causes, during the second stage, Gemba analysis was carried out. A few examples of these are illustrated in the remaining part of the analyse phase. The type of data used for analysis and the results obtained are also explained in detail.

For example, one such cause requiring validation was the dependency of failure type and the type of terrain. A chi-square test was performed for the data between type of failure and type of terrain, and the p-value was found to be 0.002 confirming that the failure type and type of terrain were dependent.

In the Gemba method of cause validation, one needs to visit the workplace and observe the practices followed and compare with the specification and/or desired method. One of the causes, 'improper compaction', is verified through this method. The requirement for compaction is that maximum dry density (MDD) should be greater than 95%. However, during process observation, it was found that in 50% of the cases (16 cases out of 32), the value is below 95%. In order to verify the effect of MDD on road defects, the number of defects is also observed in these 32 cases. A two-sample Poisson rate analysis is carried out for the data given in Table 9.11.

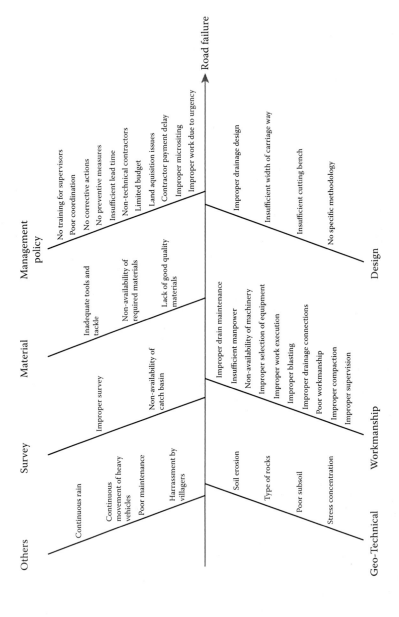

Figure 9.6 Cause and effect diagram for road failure.

Table 9.11 Details of two-sample Poisson rate test

Sample	Total occurrences	Number of defects	Rate of occurrence
MDD below 95%	15	16	0.9375
MDD above 95%	1	16	0.0625

Table 9.12 Selected solutions

Sl. No.	Root cause	Solution
1	Inappropriate methodology	Preparation of standard and uniform methodology
2	Inadequate specification	Quality checklist
3	Improper supervision	Recording of the observations at each stage with signature
4	Non-availability of required materials	Reconnaissance survey to identify the sources of materials and suggest methodology accordingly
5	Improper work due to urgency	Proper planning and review of job status on a weekly basis at site
6	Villagers' harassment	Not to give false commitment to land owners. Commitment to be recorded and circulated among the stakeholders. Develop social relationship with the villagers. Shall discuss with the competitors for uniform commitment.
7	Improper drain maintenance	Preparing a maintenance manual and protocol for checking at frequent intervals
8	Improper compaction	Checking the compaction during the construction of the road was introduced in the checklist, and the supervisors were given responsibility for execution

The test has given a Z value of 3.5 with p-value $= 0.000$. Hence 'improper compaction' is considered as a root cause.

Similarly, all the remaining causes were validated either through statistical analysis or through the Gemba analysis. Through this cause validation, a total of eight root causes were identified. (These are included in Table 9.12).

9.2.3.4   Improve phase

A brainstorming session was conducted with the team and all stakeholders of the process to identify the solutions for the selected causes.

During this discussion, solutions were identified for all eight root causes selected during the analyse phase. The selected solutions along with the root causes are presented in Table 9.12. In order to prioritise the improvement actions, a benefit–effort matrix is used, and the same is presented in Figure 9.7. It may be noted here that various methods exist to prioritise the improvement actions based on the perception of the team, and in this project scenario, a benefit–effort matrix was found to be most acceptable technique. This benefit–effort matrix identifies the actions in four categories: highly desirable actions, potentially desirable actions, potential quick hits and least desirable actions. As per the team assessment, the desirable actions are those which require less effort but result in high benefit. Based on this analysis, the actions are prioritised for implementation. A risk analysis was performed to identify the potential risk associated with the desired actions (solutions for root causes), and it was concluded that there is no risk associated with any of the solutions identified. An implementation plan was prepared with details of actions planned with responsibility and target date for completion.

A pilot run of the solutions was tried in one of the selected projects of a customer in one of the states of India. The results of the solutions implemented were observed during the pilot project and are shown in Figure 9.8. The data on defects observed were recorded during the implementation. The DPMO of the improved process was found to be 1852, resulting in an approximate sigma level of 4.40, which shows significant improvement in the process.

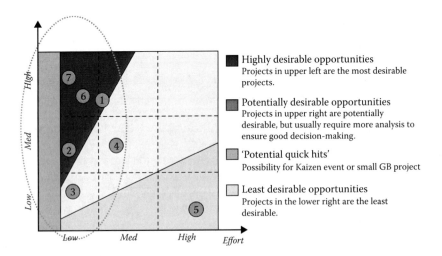

Figure 9.7 Benefit–effort matrix.

Figure 9.8 Road after the study.

9.2.3.5   Control phase

The objective of the control phase of a Six Sigma project is to ensure sustainability of the results achieved during the project. Sustainability can be achieved by standardisation of practices and monitoring the performance over time. As a first step in the control phase, the process flow charts were modified in line with the implemented solutions. This revised process flow chart was circulated to all project sites and regional offices for strict adherence to the modified procedures. A quality checklist was prepared for road construction, so that at different stages during the project execution, the engineers can ensure that all necessary steps are taken care of. A periodic audit was planned by the corporate team to ensure that all the procedures were strictly followed during the construction activities. The occurrences of 'road defects' in each project site were monitored monthly and discussed during monthly progress review meetings.

9.2.4   Managerial implications

It is amply demonstrated through the case study that a structured methodology like Six Sigma can be used effectively in road construction. This was the first time in the organisation that a checklist was prepared for activities related to road construction to control the processes. There were isolated efforts in the organisation in the past to implement initiatives like statistical process control, quality circles, small group activities, Kaizen, etc. During the implementation of those initiatives, no systematic effort was made to identify the improvement opportunities in line with business priorities or customer requirements. As a result, the impact was not very visible in the organisation, whereas in Six Sigma, projects were identified with respect to the voice of the business and the customer, and the problems addressed were of highest priority to the organisation. Hence, the management decided to use Six Sigma methodology for all future improvement initiatives in the organisation. This all may appear simple, but it is typically not easy to implement such changes in organisations/ businesses that are culturally not used to practise innovation. However, it is highly effective and can be accomplished with the right organisational infrastructure.

9.2.5   Key lessons learned from the case study

Six Sigma methodology helped the people in the organisation to understand how a process problem can be addressed systematically. During the project, extensive data collection and analysis were performed to make meaningful conclusions regarding the process. Once data collection started, hidden problems in the process were uncovered. Learning statistical software like Minitab and JMP along with Six Sigma has strengthened the ability of the people to make data-based decisions. In this process, everyone in top management and the team understood the power of data-based decision-making.

The Six Sigma methodology has helped the organisation to address some of the vital problems in procurement, land development, manufacturing, installation and servicing of windmills for its customers. Like any other business process improvement technique, improvement in road quality must be based on facts, which objectively establish the root cause(s). The Six Sigma approach for road quality improvement will not only result in more effective control over the road development process but also permit this objective to be accomplished by saving time, effort and money.

9.2.6   Recap of tools used

The tools used during this case study:

Cause and effect diagram
Chi-square test
Failure mode and effect analysis
Gemba analysis
Pareto analysis
SIPOC
Sigma level calculation
Two-sample Poisson rate test

9.2.7 Summary

The purpose of this case study is to apply Six Sigma methodology, a systematic and structured approach for quality improvement in the wind energy sector. This study has addressed one of the very critical processes in the installation and maintenance of windmills. The quality of wind farm roads is a major problem encountered by all windmill manufacturers in India. This is one of the first case studies of Six Sigma in use in this field in India, and it was successfully completed. As a result of this study, the failure of roads was reduced significantly. This project has resulted in a direct saving of US$168,000 for the company per annum. This includes the cost reduction in repair of damaged roads and the cost of equipment waiting time at various sites across the country. The cost of rental equipment used for repairs to the road is very high, in addition to the waiting time of trailers loaded with windmill parts/spares to be installed/repaired. This study also has helped the organisation to complete the installation activities on time, which has resulted in improvement in customer satisfaction. This project has further helped the organisation to ensure availability of spare parts for maintenance activities. In total, this project has had a significant impact in all the field activities of the company. The results of this study provide greater stimulus for the wider application of Six Sigma methodology across the company in the future. Also, this case study demonstrates the applicability of Six Sigma methodology in a novel field.

9.3 Case study 3: Application of Six Sigma methodology to reduce the rejection and rework in an automobile supplier company

9.3.1 Background of the company

The company under study is a small-scale organisation catering to a large automobile supplier company. This company has around 150 employees, manufacturing the components/parts used for automobiles. As the requirement for quality improvement was very stringent for the suppliers of automobile companies, this organisation used to practice a few quality improvement techniques such as statistical process control, FMEA, etc. After implementing such initiatives in the process, rejection and rework were still a big problem for the organisation.

9.3.2 Background to the problem

This case study deals with reduction of rejection and rework of hinge hole diameter of the honing process in an automobile part manufacturing company. This component is the critical connecting link of flyweight assembly used in fuel injection pumps. Flyweight assembly in turn is a part of governor, which regulates the quantity of fuel injected into the engine, and thereby regulates the engine speed. If the hinge hole is undersized, it can cause undue stresses in the connecting link. This may lead to cracks in the link and results in failure of the pump. If the hinge hole is oversized, it may make the flywheel sticky, leading to failure of the governor function. The first pass yield of this process was only 87.8%.

9.3.3　Six Sigma methodology (DMAIC)

As the first pass yield was very low, it was affecting on-time delivery of the components to the customer. Also, the cost of rejection and rework was very high. Solving this problem was very critical to the management of the company as it was clear that an effective solution to this problem would have a significant impact in reducing rework/rejection and improving customer satisfaction. Hence, it was decided to address this problem by the application of Six Sigma methodology (Gijo and Scaria 2010). In this case, since the existing process requires improvement, the DMAIC methodology was applied. The remaining part of this section presents various phases of Six Sigma DMAIC methodology.

9.3.3.1　Define phase

The define phase of the Six Sigma methodology aims to define the improvement project in terms of customer requirements and identify the underlying process that needs improvement. The first step was to develop a project charter with all the necessary details of the project, including team composition and schedule for the project. The project charter is provided in Table 9.13. This has helped the team members clearly understand the project objective, project duration, resources, roles and responsibilities of team members, project scope and boundaries, expected results from the project, etc. This creates a common vision and a sense of ownership for the project, so that the entire team is focused on the objectives of the project. The project team included a champion, a black belt, three green belts and three operators from the process. During the define phase of the project, the team, along with the champion, had detailed discussions regarding the problem. The project team defined the goal statement of the project as improving the yield of the honing process from the current level of 88% to 98%, which should result in significant reduction in rejection and rework.

A basic flow chart of the process was prepared, and a SIPOC mapping was carried out to have a clear understanding of the process. The team focused on the honing process for improvement, which is defined as the scope of the project. The process mapping along with SIPOC (refer Table 9.14) provides a picture of the steps needed to create the output of the process.

Table 9.13 Project charter

Project Title· Reducing the rejection and rework in the honing process.

Background and reason for selecting the project:

The problem is complex, and there are too many variables affecting the tolerance. We were unsuccessful in finding the solution in the previous attempts. The first pass yield of the honing process is only 87.8% for the past 6 months. Out of 12.2% defective components, close to 1.3% components were scrapped, and the remaining were reworked. Thus the cost of repair and scrap was very high and led to the delay in delivery of components to the customer.

Aim of the project: To improve the first pass yield from 88% to 98%.

Project champion	Head—Production Department
Project leader	Manager—Production
Team members	Engineer—Production, Quality Control Inspector Supervisor—Maintenance Operator—Shift I, Operator—Shift II, Operator—Shift III

Characteristics of product/process output and its measure

CTQ	Measure & specification	Defect definition
Diameter	9.000–9.009 mm	Diameter beyond 9.000–9.009 mm
Expected benefits	Reduction in rework and scrap as a result of reduction in diameter variation. This will help the organisation to improve on-time delivery of components to its customers.	
Schedule	Define: 2 weeks Measure: 2 weeks Analyse: 4 weeks Improve: 4 weeks Control: 4 weeks	

Table 9.14 SIPOC

Supplier	Input	Process	Output	Customer
Heat treatment department	Component	Honing process	Machined component	Assembly shop
Metrology department	Preprocess gages		Production report	Production department
Engineering department	Instruction charts			

Honing process

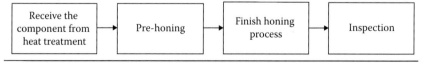

9.3.3.2   Measure phase

This phase is concerned with selecting appropriate product characteristics, mapping the respective processes, studying the accuracy of measurement system, making necessary measurements, recording the data and establishing a baseline of the process capability or sigma rating for the process.

In this project, the CTQ characteristic considered for further study is the hinge diameter. The specification limit for the diameter is from 9.000 mm to 9.009 mm. Since the tolerance is only nine microns, it was necessary to validate the measurement system by conducting a gage R&R study. For conducting this study, three operators working with this process were identified, along with 10 components. After collecting the data, analysis was performed, and the total gage R&R value was found to be 14.83%. Since this value is within the acceptable limit of 30%, it was concluded that the measurement system is acceptable for further data collection. Then a data collection plan was prepared with details about characteristics for which data were collected, including sample size and frequency of data collection with details of stratification factors. As per the data collection plan, data were collected for diameter, and the same is tested for normality by the 'Anderson–Darling normality test' with the help of Minitab statistical software. From the Minitab software output (Figure 9.9), the p-value was found to be less than 0.05, which leads to the conclusion that the data are from a population that is not normal. Hence, from the observed performance of the process capability analysis from

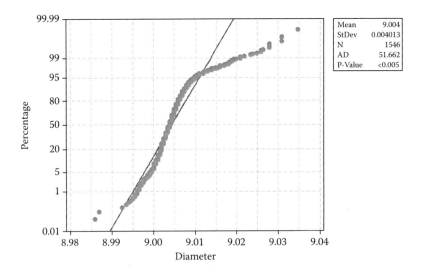

Figure 9.9 Normal probability plot for diameter.

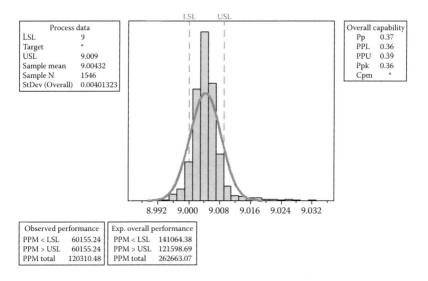

Figure 9.10 Process capability of diameter.

Minitab output (Figure 9.10), the parts/million (ppm) total was identified as 120,310, and the corresponding sigma level was found to be 2.67.

9.3.3.3   Analyse phase

The objective of the analyse phase in this study is to identify the root cause(s) that create the dimensional variation during the honing process. Hence, in the analyse phase, a brainstorming session was conducted with all the team members and associated personnel to identify the potential causes for dimensional variation. The causes identified during the brainstorming session were presented as a cause and effect diagram (Figure 9.11). For validating the causes, the type of data possible to collect on each of these causes was identified. Based on the availability of the data on the causes, a decision was taken about the type of analysis possible to validate each one of these causes. It was found that some of these causes can only be validated by Gemba, and different types of statistical analysis can be performed on data collected for the remaining causes. Based on this understanding, a cause validation plan was prepared for all the potential causes and is presented in Table 9.15. This cause validation plan gives the details of analysis planned for causes. For those causes where Gemba was identified as the method of validation, the team observed the process at random frequency for a period of 1 month, the observations were noted and a decision was made over whether it is a root cause or not. The details of analysis and validation of the causes are presented below.

Data were collected on input and process parameters like input part size, spindle-1 feed, spindle-2 feed, spindle impulse-1, spindle impulse-2,

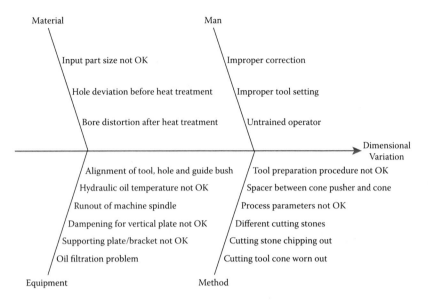

Figure 9.11 Cause and effect diagram for dimensional variation.

Table 9.15 Cause validation plan

S. no.	Causes	Validation method
1	Input part size not OK	Regression analysis/DoE
2	Hole deviation before heat treatment	Gemba
3	Bore distortion after heat treatment	Gemba
4	Cutting tool cone worn out	Gemba
5	Cutting stone chipping out	Gemba
6	Different cutting stones	Gemba
7	Process parameters not OK	Regression analysis/DoE
8	Spacer between cone pusher and cone	Gemba
9	Tool preparation procedure not OK	Gemba
10	Oil filtration problem	Gemba
11	Supporting plate/bracket not OK	Gemba
12	Dampening for vertical plate not OK	Gemba
13	Runout of machine spindle	Gemba
14	Hydraulic oil temperature not OK	Regression analysis/DoE
15	Alignment of tool, hole and guide bush not OK	DoE
16	Improper correction	Gemba
17	Improper tool setting	DoE
18	Untrained operator	GEMBA

output spindle size and hydraulic oil temperature with corresponding diameter. The effects of the input and process parameters on the diameter variation were validated by a regression analysis.

Before performing the multiple regression analysis, the variables were tested for 'multicollinearity'. From the 'variance inflation factor' (VIF) of the regression analysis (Tables 9.16 and 9.17), it is evident that 'multicollinearity' is not present in the data. Since the p-values for spindle-2 feed, spindle impulse-2, output spindle size and hydraulic oil temperature from the regression analysis were found to be less than 0.05, it was concluded that these variables significantly affect the diameter variation.

The process parameters like feed rate, impulse, stroke, etc., were earlier fixed based on the experience of the operators. Hence, it was decided to conduct a DoE during the improvement phase to identify the optimum process setting for these parameters. The other causes listed in the cause and effect diagram were validated by Gemba analysis. The detail of validation of all causes is summarised in a tabular format and is given in Table 9.18.

9.3.3.4   Improve phase

As per the decision of the team in the analyse phase, a DoE is planned during this phase. The parameters selected for experimentation are 'feed rate', 'impulse', 'stock', 'stroke' and 'oil temperature'. It was also felt by the

Table 9.16 Minitab software output of regression analysis

Predictor	Coefficient	SE of coefficient	t-statistic	p-value	VIF
Constant	6.682	0.6297	10.61	0.000	–
Input part size	0.00429	0.01082	0.40	0.692	1.1
Spindle—1 feed	0.0000728	0.000142	0.51	0.609	1.4
Spindle—2 feed	0.0005951	0.0002792	2.13	0.035	1.6
Spindle impulse—1	0.00000038	0.0000525	0.01	0.994	1.4
Spindle impulse—2	0.00150339	0.00006605	22.76	0.000	1.7
Output spindle size	0.24731	0.06792	3.64	0.000	1.6
Hydraulic oil temperature	–0.00005682	0.00001826	–3.11	0.002	1.5

Table 9.17 ANOVA table for regression analysis

Source	DF	SS	MS	F	p-value
Regression	7	0.00152454	0.00021779	132.58	0.000
Residual error	131	0.00021520	0.00000164		
Total	138	0.00173974			

Table 9.18 Validated causes

S. no.	Causes	Conclusion
1	Input part size not OK	Not a root cause
2	Hole deviation before heat treatment	Not a root cause
3	Bore distortion after heat treatment	Not a root cause
4	Cutting tool cone worn out	Not a root cause
5	Cutting stone chipping out	Not a root cause
6	Different cutting stones	Not a root cause
7	Process parameters not OK	Root cause/to be optimised by DoE
8	Spacer between cone pusher and cone	Not a root cause
9	Tool preparation procedure not OK	Not a root cause
10	Oil filtration problem	Not a root cause
11	Supporting plate/bracket not OK	Not a root cause
12	Dampening for vertical plate not OK	Not a root cause
13	Runout of machine spindle	Not a root cause
14	Hydraulic oil temperature not OK	Root cause/to be optimised by DoE
15	Alignment of tool, hole and guide bush not OK	Root cause/to be optimised by DoE
16	Improper correction	Not a root cause
17	Improper tool setting	Root cause/to be optimised by DoE
18	Untrained operator	Not a root cause

team that there might be possible interaction of 'feed rate' with 'impulse' and 'stock'. Hence, it was also decided to estimate the effect of these two interactions. Since the relationship between these variables and the diameter is not established as linear, all these factors were experimented at three levels. The existing values for the parameters were considered as one level for experiment. Five factors at three levels and two interactions require a huge number of components for conducting full factorial experiments. Hence, it was decided to use $L_{27}(3^{13})$ OA for conducting this experiment. A master plan for the experiment (Table 9.19) was prepared by allocating all the experimental factors in $L_{27}(3^{13})$ OA. The experimental sequence given in the master plan was randomised, and experimentation was done. For each experiment, the diameter was measured. These data were analysed by Taguchi's S/N ratio method. Since the diameter is nominal, the best type of characteristic, the S/N ratio formula used for analysis was $10 \log(\bar{Y}^2/s^2)$, where $\bar{Y}$ is the average and s the standard deviation for each experiment. The main effect plots and interaction plots were made

Table 9.19 Master plan for experimentation with the collected data

Exp. no.	Feed rate	Impulse	Stock	Stroke	Temperature	Diameter		
						1	2	3
1	1	32	30	18	30	9.00500	9.00375	9.00300
2	1	32	50	20	40	9.00600	9.00525	9.00550
3	1	32	70	22	50	9.00300	9.00175	9.00425
4	1	35	30	20	40	9.00525	9.00325	9.00350
5	1	35	50	22	50	9.00425	9.00400	9.00350
6	1	35	70	18	30	9.01050	9.00700	9.00825
7	1	38	30	22	50	9.00650	9.00625	9.00550
8	1	38	50	18	30	9.00675	9.00550	9.00600
9	1	38	70	20	40	9.00300	9.00375	9.00400
10	2	32	30	18	40	9.00475	9.00425	9.00425
11	2	32	50	20	50	9.00425	9.00325	9.00250
12	2	32	70	22	30	9.00400	9.00400	9.00325
13	2	35	30	20	50	9.00350	9.00425	9.00300
14	2	35	50	22	30	9.00125	9.00400	9.00300
15	2	35	70	18	40	9.00400	9.00250	9.00175
16	2	38	30	22	30	9.00425	9.00400	9.00275
17	2	38	50	18	40	9.00475	9.00500	9.00450
18	2	38	70	20	50	9.00450	9.00500	9.00475
19	3	32	30	18	50	8.99625	8.99275	8.99675
20	3	32	50	20	30	9.00350	9.00525	9.00500
21	3	32	70	22	40	9.00325	9.00200	9.00100
22	3	35	30	20	30	9.00400	9.00250	9.00600
23	3	35	50	22	40	9.00525	9.00500	9.00275
24	3	35	70	18	50	9.00375	9.00500	9.00625
25	3	38	30	22	40	9.00150	9.00300	9.00500
26	3	38	50	18	50	9.00200	9.00400	9.00400
27	3	38	70	20	30	9.00300	9.00325	9.00300

for the S/N ratio values (Figures 9.12 and 9.13). The level that maximises the S/N ratio was selected as the best level for that factor. Thus, the best levels for the factors were identified from the main effect plots and interaction plots. The optimum factor level combination identified is presented in Table 9.20.

These optimum levels identified were considered as the solution for these process parameters. The team had detailed discussions involving all stakeholders of the process, and solutions were identified for all the remaining root causes. The solutions identified are presented in Table 9.21. A risk analysis was conducted for identifying possible negative side effects of the solutions during implementation. The team has

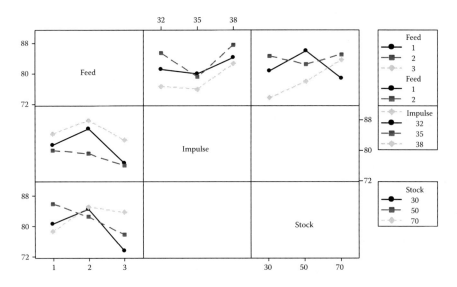

Figure 9.12 Interaction plot (data means) for S/N ratios.

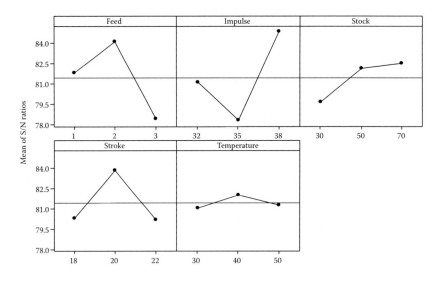

Figure 9.13 Main effects plot (data means) for S/N ratios.

concluded from the risk analysis that there is no risk associated with any of the identified solutions. After the risk analysis, an implementation plan was prepared for all solutions with responsibility and target dates for completion for each solution. The solutions were implemented as per the plan, and results were observed. The data on diameter were collected

Table 9.20 Optimum factor level combination

S. no.	Factor	Optimum level
1	Feed rate in microns	2
2	Impulse in microns	38
3	Stock in microns	70
4	Stroke in mm	20
5	Hydraulic oil temperature in degrees	40

Table 9.21 Validated causes and solutions

S. no.	Validated cause	Solution
1	Process parameters not OK	Established optimum parameters by DoE
2	Hydraulic oil temperature not OK	Established optimum parameters by DoE
3	Alignment of tool, hole and guide bush not OK	1 Introduced Poka-Yoke system to ensure alignment 2 Alignment between headstock and tailstock with respect to axis corrected. This is introduced as a parameter to be checked in machine preventive maintenance checklist.
4	Improper tool setting	Optimised the setting parameters by DoE

from the process after the project. The process capability evaluation was done, and the details are provided in Figure 9.14. The PPM level of the process was 0, and the corresponding sigma rating was 6 (Table 9.22). A dot plot (Figure 9.15) was made for comparing the process before and after the project, which shows significant reduction in dimensional variation after the project.

9.3.3.5   Control phase

The real challenge of Six Sigma implementation is not in making improvements in the process but in sustaining the achieved results. Due to many organisational reasons like people changing jobs, maintaining

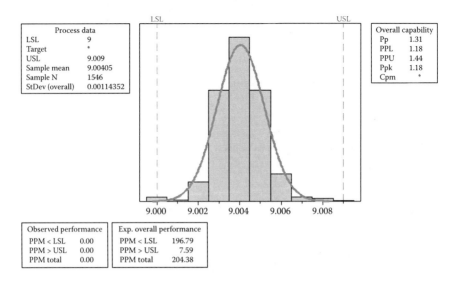

Figure 9.14 Process capability after the project.

Table 9.22 Comparison of results
before and after

	Before	After
Sigma level	2.67	6.0
DPMO	120310	0
Yield	88%	100%

the results is often extremely difficult. Standardisation of the improved methods and continuous monitoring of the results only can ensure sustainability of the results. It is also important to ensure that the operating personnel in the process feel a sense of ownership in the solutions implemented, so that without any external intervention the process can be maintained.

Since the organisation was implementing the ISO 9001:2008 QMS, the process changes were documented in the procedures of the QMS. This has helped to standardise the improved methods in this project. A 'run chart' was introduced for monitoring the process, along with a reaction plan. This reaction plan helps the operators to take action on the process in case assignable causes occur. Training was provided to the people working with the process about the improved operational

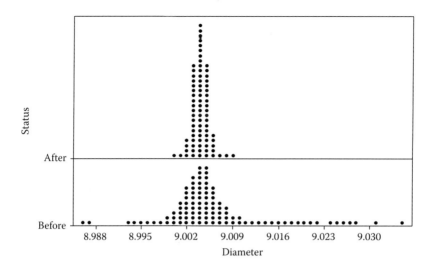

Figure 9.15 Dot plot of diameter before and after the project.

methods so that their confidence level in working with the new process increased.

9.3.4   *Managerial implications*

This case study was an eye opener for the management as it delivered a significant improvement in the process. Data and their analysis gave confidence to the people and top management for making decisions about the process. The success in this project has made them the 'change agents' in the process of cultural transformation of the organisation. Management introduced a team known as the 'leadership team' in the organisation to oversee the Six Sigma project selection and execution. All issues related to implementation were also reported to this team for further action.

9.3.5   *Key lessons learned from the case study*

The main learning points from the case study can be stated as follows. The Six Sigma exercise provided the company with an example of the benefits of addressing a problem systematically. Extensive data collection was essential to the success of the project, but this had to be focused on the key areas identified in the study. Also, no amount of data collection would be valid without the gage R&R study. Statistical software was essential for the analysis. However, these packages require use by people with the correct training. Management and staff began to believe in their own ability to implement advanced methods. The good example set by this project, supported by making Six Sigma a factor in staff appraisal, has encouraged staff to accept the use of the technique. Over the time period of the project, difficulties from loss of trained staff delayed the project. Future projects would benefit from training additional staff beyond initial requirements.

9.3.6   Recap of tools used

The tools used during this case study:

Analysis of variance
Cause and effect diagram
Design of experiments
Dot plot
Gemba analysis
Main effect plot
Normality test
Orthogonal array
Process capability analysis
Regression analysis
SIPOC
Sigma level calculation

9.3.7   Summary

As a result of this study, the first pass yield has improved from 88% to 100%. The sigma rating of the process showed improvement from 2.67 to 6.0 after the project. The team, with the help of the finance department, estimated the tangible savings of this project. It was found that the cost associated with scrap, repair and tool has come down drastically. This has given encouragement to management to implement Six Sigma methodology for all improvement initiatives in the organisation. To encourage the people in the organisation to use Six Sigma methodology, management decided to suitably reward the successful teams. After observing the success in this project, the people were more confident in implementing Six Sigma in addressing any improvement initiative in the organisation.

9.4 Case study 4: Application of value stream mapping in a camshaft manufacturing organisation

9.4.1 Background of the company

The company where the case study was performed started 20 years ago as a small-scale company which manufactures camshafts. The company started with 20 employees, and current employee strength is 60. The current turnover of the company is 4,000,000 Indian rupees (US$63,000). The company had no scientific implementation of Lean initiatives before. Certainly some remote activities of Lean manufacturing were prevailing in the organisation.

9.4.2 Background of the problem

The camshafts were in high demand compared to other products manufactured by the company. A macro level analysis on the camshaft manufacturing line implied that wastes occur in manufacturing, and there is potential to improve value addition of the manufacturing line. Hence, it has been decided to apply value stream mapping (VSM) to the camshaft manufacturing line to identify and analyse wastes and improve value addition.

9.4.3 Value stream mapping methodology

VSM is one of the powerful Lean tools used to identify wastes and detect opportunities for value addition. It starts with analysing the current conditions and developing a current state map, identifying improvement opportunities and developing a future state map which indicates the desired state of performance. The development stages of VSM are detailed in the remaining parts of this section.

9.4.3.1 Formation of task force
A cross-functional task force team was formed with experts from design, manufacturing and quality control and academic researchers. The task force analysed the manufacturing stream using VSM methodology and identified actions for improvement.

9.4.3.2 Process map and data collection
The task force prepared the process map shown in Figure 9.16. The process map indicates the process sequence for the manufacturing of camshafts. The data required for each process are collected using the checklist format

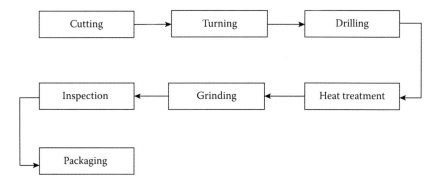

Figure 9.16 Process map for camshaft manufacturing line.

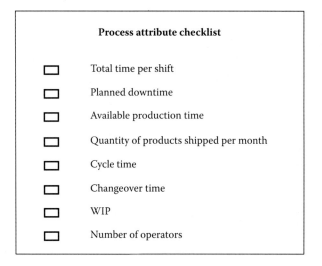

Figure 9.17 Process attributes checklist.

shown in Figure 9.17. Process activity mapping is performed to capture the overall view of processes involved in manufacturing the product. It gives details about product flow and operations and details of manpower involved in manufacturing the product. The process activity map for the manufacturing line is shown in Figure 9.16.

The task force started collecting the necessary data by visiting the shop floor, and the collected information is shown in the data collection sheet (refer to Figure 9.18).

Customer requirements

- Average demand = 100 units

Supplier information

XYZ receives shipment of 1,200 units from its supplier

Available production time = 430 min

Cutting

Cycle time = 15 min

Changeover time = 4 min

Uptime = 99.06%

Turning

Cycle time = 15 min

Changeover time = 5 min

Uptime = 99.06%

Drilling

Cycle time = 20 min

Changeover time = 5 min

Uptime = 99.06%

Heat treatment

Cycle time = 4 min

Changeover time = 4 min

Uptime = 99.06%

Grinding

Cycle time = 5 min

Changeover time = 3 min

Uptime = 99.1%

Inspection

Cycle time = 4 min

Changeover time = 2 min

Uptime = 99.3%

Packaging

Cycle time = 3 min

Changeover time = 0 min

Uptime = 100%

Flow of material and information

All communications with customer and supplier are electronic

Production control getsmonthly orders from customers

Production control generatesdaily orders to process owners

Figure 9.18 Data collection sheet.

9.4.3.3   *Total product cycle time*

- It is the actual time taken to complete a task and proceed to the next step.
- It is the time taken to perform the corresponding process.
- It is the time that elapses between one part coming off the process and the next part going in.

- One of the vital goals of Lean is to match cycle time with takt time.
- The individual cycle time and total cycle time for manufacturing the camshaft are shown below:

	minutes
Cutting	15
Turning	15
Drilling	20
Heat treatment	4
Grinding	5
Inspection	4
Packaging	3
Total value stream cycle time	66

9.4.3.4   Takt time

Takt time is defined as average unit production time required to fulfil customer demand. The production cycle time must always be less than takt time to maintain flow in the manufacturing process. It is calculated as the ratio of net available time to daily demand. It is denoted in minutes. The takt time for manufacturing line was calculated and found to be 4.30 minutes.

$$\text{Takt time} = \left(\frac{\text{Net Available Time}}{\text{Daily Demand}} \right) = \frac{430}{100} = 4.30 \text{ min}$$

9.4.3.5   Analysis of Lean metrics

Total value stream WIP inventory

Raw material prior to cutting	1200 units
Between cutting and turning	5 units
Between turning and drilling	5 units
Between drilling and heat treatment	5 units
Between heat treatment and grinding	5 units
Between grinding and inspection	20 units
Between inspection and packaging	120 units
Finished product after packaging	150 units
Total inventory	1510 units

9.4.3.6   WIP calculation

Work-in-process (WIP) inventory denotes inventory that has been partly converted through the production process, and extra work must be finished before it can be transported out of the manufacturing line and depicted as finished goods inventory.

Raw material prior to cutting	12 days on hand
Between cutting and turning	0.05 days on hand
Between turning and drilling	0.05 days on hand
Between drilling and heat treatment	0.05 days on hand
Between heat treatment and grinding	0.05 days on hand
Between grinding and inspection	0.2 days on hand
Between inspection and packaging	1.2 days on hand
Finished product after packaging	1 days on hand
Total inventory	14.6 days on hand

9.4.4   Bottleneck analysis

A bottleneck process is defined as the process which stops or slows the flow of the manufacturing process. The stoppage can be due to delay in material arrival, unavailability of manpower, not following standard operating procedures, etc. In a manufacturing line, a bottleneck station can be identified as the station which has its cycle time greater than its takt time. In the camshaft manufacturing line based on the process individual cycle time and total takt time, bottleneck analysis was performed, and bottleneck processes were identified. Figure 9.19 shows the bottleneck stations for the camshaft manufacturing line.

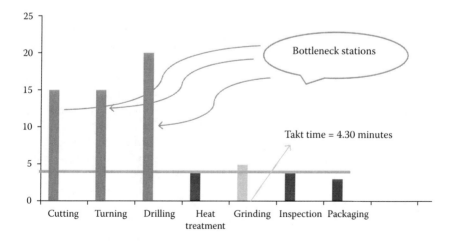

Figure 9.19 Bottleneck analysis for the manufacturing line.

9.4.5   Description of current state map

The current state map captures the present condition of the manufacturing line. The purpose of current state VSM is to identify value-added and non-value-added activities by tracking information and material flow within the manufacturing line. The manufacturing line where the study is conducted manufactures camshafts. Cutting is the first operation to be performed on raw material, followed by turning and drilling operations. Later the machined component is heat treated and finally, grinding is performed. Next, the manufactured camshaft is inspected and sent for packaging and dispatching. A total of 11 workers are involved in manufacturing the camshafts, and their tasks are properly assigned. The firm operates for 8 hours/day (including a 30-minute lunch break and 10-minute tea break) and manufactures 100 camshafts/day. The individual cycle time for each process was found out by conducting time study, and WIP inventory data were also collected visually. Based on the available data, total cycle time and lead time was calculated and was found to be 66 minutes and 14.6 days. Takt time was found to be 4.3 minutes, and VA ratio for the current state was 0.31%. After analysing the current state map, potential improvement actions were planned and implemented to improve the current state. The current state map for the manufacturing line is shown in Figure 9.20.

Total cycle time	66 min
Process lead time	$14.6\,\text{days} = 14.6 \times 24 \times 60 = 21{,}024\,\text{min}$
VA ratio	$66/21{,}024 \quad = 0.31\%$

9.4.6   Improvements and future state map

The total time required to manufacture the camshaft was 66 minutes. The process lead time was calculated and found to be 14.6 days. Takt time was calculated as 4.30 minutes. VA analysis was performed to find out the bottleneck stations. Based on VA analysis, the processes cutting, turning and drilling were identified as the bottleneck stations. As a part of improvement activity, 5S (seiri, seiton, seiso, seiketsu and shitsuke) was performed on the entire manufacturing line to create an ordered workplace. Further tool kits and trolleys were provided to reduce the time spent on searching for tools and for transportation. Apart from these improvements, minor Kaizen activities were also planned and implemented, and results were observed.

9.4.6.1   Improvements pertaining to cutting process

The cutting process utilises a cutting machine for sizing the raw material needed for processing from the bar stock. The machine uses a sawtooth

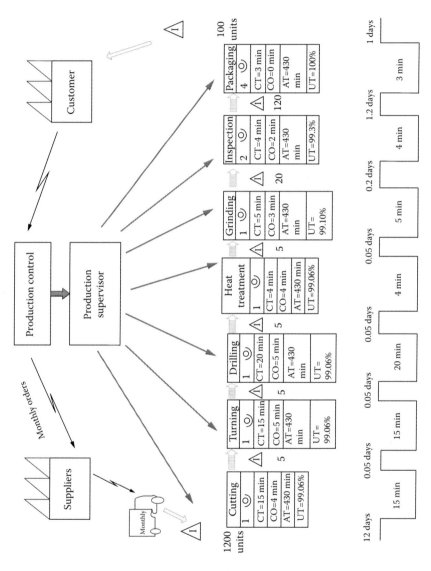

Figure 9.20 Current state map for camshaft manufacturing line.

blade as the cutting tool for sizing. After analysing the performance of the cutting tool, the old tool was replaced with a new tool of greater thickness which is capable of operating at higher cutting speeds. A three-jaw holding device was installed to reduce the time spent on loading and unloading. Further space was provided to store the accessories such as grease waste cloth and spanners to avoid unnecessary movement. All these actions improved the process efficiency, and cycle time was brought down to 11 minutes from 15 minutes.

9.4.6.2 Improvements pertaining to turning and drilling processes

The turning operation is performed with a centre lathe. It uses a four-jaw chuck for holding the job, and more time is spent on loading and unloading the job. The four-jaw chuck was replaced with a three-jaw chuck to simplify the process of loading and unloading as it requires less effort and saves time. The currently used single point cutting tool was replaced with an insert based single point cutting tool, which helped in achieving better accuracy and process speed. Further quick changeover and single minute exchange of die (SMED) concepts were applied apart from 5S and minor Kaizen activities to reduce the process cycle time. After implementing the improvement actions, the cycle time was found to be 10 minutes.

The drilling operation is performed with a special purpose radial drilling machine. Studies were conducted to improve spindle speed and feed rate of the drilling machine, and based on the findings, the improvement actions were implemented. Stack height and retract rate of the drilling machine were synchronised to improve the drilling efficiency at greater speeds. Further jigs and fixtures were provided to reduce changeover time and time spent on loading and unloading. SMED and Poka-Yoke (mistake proofing or foolproofing) concepts were deployed while designing jigs and fixtures. The cycle time was brought down to 15 minutes from 20 minutes after implementing the improvement actions. Apart from the discussed improvements, maintenance and training programs were also conducted to identify and implement the improvement actions. The future state map for the manufacturing line is shown in Figure 9.21.

Total cycle time	48 min
Process lead time	8.2 days = 8.2 × 24 × 60 = 11,808 min
VA ratio	48/11,808 = 0.41%

The VA ratio has increased from 0.31% to 0.41%, which shows a 32% improvement.

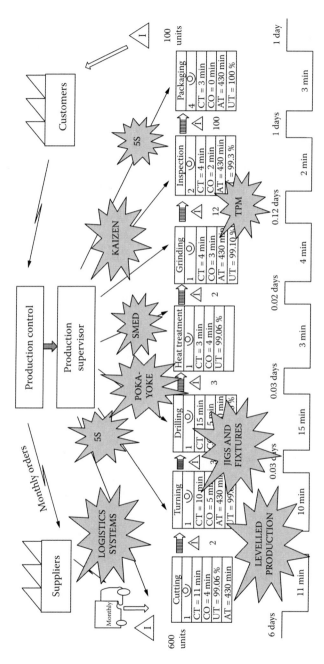

Figure 9.21 Future state map for camshaft manufacturing line.

9.4.7   Comparison of current and future state maps

After implementing the improvement actions, a comparison was made to measure the Lean metrics before and after implementation. Table 9.23 shows the comparison of the Lean metrics before and after implementation. The total cycle time was reduced from 66 minutes to 48 minutes. Total value stream WIP inventory was reduced from 2310 units to 1121 units. The value stream lead time was found to be 8.22 days after implementation. Finally, the VA ratio was improved from 0.31% to 0.41%.

9.4.8   Managerial implications

The case study has enabled senior management to make effective decisions on Lean implementation. Prior to the conduct of the study, no systematic and unified efforts were taken to implement Lean tools, and improvements were not realised. The results of the present study initiated Lean culture in the organisation and mindset transformation among the workforce. During this study, a task force team was constituted with all divisional heads of the case organisation. The task force members were trained on Lean concepts. The task force members analysed the current state and identified the improvement initiatives using the primary Lean tool called VSM. The identified initiatives were subjected to implementation for achieving process excellence, streamlined production and quality enhancement. The conduct of the pilot study enabled the inculcation of Lean culture in the organisation.

9.4.9   Summary

This case study presents the application of VSM to a camshaft manufacturing process. A task force team was formed with members from

Table 9.23 Comparison of metrics before and after VSM implementation

Metrics	Before implementation	After implementation	Percentage improvement (%)
Total cycle time	66 min	48 min	27% reduction
Total value stream WIP inventory	2310	1121	51% reduction
Value stream lead time	14.6 days	8.22 days	43% reduction
VA ratio	0.31%	0.41%	32% improvement
Problem solving competence	Less problem solving skills	Enhanced problem-solving skills	
Team morale	Low	High	

different divisions. The task force members analysed the current status of value stream after collecting the data using an attribute checklist. Based on analysis of current status, the desired future state was developed by achieving streamlined production of the camshaft. The improvements in terms of cycle time reduction, lead time reduction, combining processes and reduction of workforce are being quantified.

9.5   Case study 5: An application of Lean Six Sigma to a die-casting process

9.5.1   Background of the company

The die-casting unit under study was established in 1978 with 150 employees, which comes under the category of SME. The organisation is engaged in designing and manufacturing various types of precision machined components using pressure and gravity die-casting processes. The main customers of the company are ordinance factories, the automobile industry and textile machine manufacturers. The company manufactures around 250,000 units of die-casting products per year to cater to the needs of its customers. The employees work in three shifts per day, each shift of 8 hours, and 6 days a week to meet the market demand.

9.5.2   Background to the problem

The die-casting process starts with placing al-alloy ingots in the furnace and heating them for a sufficient duration. When the metal melts and achieves a suitable temperature in the casting furnace, it is inserted into the dies by plunger pressure. As the metal solidifies the cast product is taken out with the help of an ejector pin and placed in a trolley. The cast product then goes to the trimming and fettling shop where extra projections are removed. The trimmed product is moved to the drilling section where the different holes and grooves are made as per the dimensions in the drawing. In the next step, semi-finished products go to the de-burring unit where the external and internal holes are cleaned and burrs are removed. The product is then moved to the chamfering and threading unit where fine cutting at different angles along the surface and the making of external and internal threads are performed. Cleaning and polishing operations are performed subsequently in the next stage. Finally, the finished product is stored in the dispatch department from where it is sent to the customer according to an agreed schedule. Customer orders are taken care of on the basis of first come first serve. Quick turnaround orders are taken care of by rescheduling the batch processing as decided by the production manager (Kumar et al. 2006).

The wish to maximise ROI and the fear of not meeting the customer demand compelled management to concentrate more on production than on quality of the finished product. This resulted in an increase in WIP inventory, scrap and rework cost, and more defects (external and internal casting defects like foliation, cracks, cold shut, pinhole porosity, etc.) in the final product. There were many hidden wastes embedded in the manufacturing process that were ignored by the company because their manufacturing capacity was higher than their production requirements. Problems were tackled by increasing WIP inventory, leading to higher inventory carrying cost. In the last 6 years, demand for their product became high due to globalisation and the boom in the automobile sector. In order to meet the customers' demand, production of automobile accessories was given top priority, irrespective of the quality of product. The management was able to meet the customer demands by putting the quality of product at risk. This resulted in a number of customer complaints from different parts of the country.

As most of the customer complaints were related to crack propagation in the final die-casting product (resulting in improper functioning of the automobile engine), management formed a team to identify the root causes of problems. Moreover, there was a constant increase in in-process inventory, machine downtime and idle time at different workstations, and there was also concern about health and safety issues of the employees as the average number of accidents on the shop floor was increasing each year.

One of the questions raised during brainstorming was related to the selection of a continuous improvement methodology from a range of existing quality improvement programmes. The team decided to implement the Lean Sigma methodology to eliminate defects, reduce variation and reduce inventory and overall complexity from the system. While Lean streamlines processes and eliminates waste (idle time, machine downtime, in-process inventory), reduces overall complexity and helps to uncover the value-added activities of a process, Six Sigma can solve complex cross-functional problems where the root causes of a problem (in this case, crack propagation) are unknown and help to reduce undesirable variations in processes. The integration of two approaches eliminates the limitations of the individual approach.

9.5.3   Lean Six Sigma methodology (DMAIC)

Although the team was using an integrated Lean Six Sigma (LSS) approach, they decided to follow the standard Six Sigma methodology (DMAIC) and to use Lean tools within the DMAIC methodology.

9.5.3.1   Define phase

A cross-functional team was formed consisting of the operators, engineers from production and quality control, the marketing department and senior managers. This team spent many hours on the shop floor observing, in order to collect data and understand the different processes associated with the die-casting unit. A number of brainstorming sessions of team members were conducted to identify CTQ characteristics based on the voice of the customer (VOC) input. In the meeting, the problem of the die-casting unit, the size of the problem, the impact of the problem, etc., were discussed among the team members, and it was apparent that most of the customer complaints related to crack propagation in the automobile accessories manufactured by the company. The goal of the team members was to identify the root cause of the problem and reduce the number of defects which occur in the product.

9.5.3.2 Measure phase

The team was divided into small groups to monitor the defects occurring in each process involved in the manufacturing of the die-casting product. The data were collected and analysed and were found to match with the historic data, showing that the maximum number of defects were coming from the die-casting machine, de-burring operation and chamfering and threading operation.

The next step was to determine a performance standard based on customer requirements. A data collection plan was established to focus on the project output and to carry out the standard setting exercise for the same. A gage R&R study was conducted to identify the sources of variation in the measurement system and to determine whether it was accurate or not. A study was performed to check the accuracy of gages used for the measurement of characteristics as well as the reproducibility of the worker in performing operations on the machine. The gage R&R study performed on the system showed a variation of 8.01%, which implied that the measurement system was acceptable. What the customers want is a sound casting with measurable characteristics, such as the density of the casting. Therefore, the ultimate goal of the team was to increase casting density.

The company was operating at a baseline capability of 0.12 with defects per unit (DPU) being 0.18. The desired specification limit of casting density was 2.73–2.78 g/cc, and the casting produced before the implementation of Lean Sigma had an average density of 2.45 g/cc.

9.5.3.3 Analyse phase

The objective of the team members was to determine the root causes of defects and identify the significant process parameters causing the defects. Out of seven casting defects, air inclusion, shrink holes, gas holes and porosity are internal defects whereas cold shut, foliations, and soldering are surface defects (external defects). The internal defects are formed during the casting process as the metal solidifies. The micro holes created inside the casting are due to air or gas entrapment and result in crack propagation due to differential pressure and force created inside the casting. This crack propagation impedes the proper functioning of the final product and thus is very significant to overall performance of the machinery where die-casting parts are fitted.

The Pareto chart shown in Figure 9.22 illustrates the percentage contribution of internal and external defects in the process. It can be concluded from Figure 9.22 that internal defects are the result of poor casting density and amount to 67% of total defects in the process. Other defects occur in the de-burring, chamfering and threading operations due to tooling and clamping problems. All the defects mentioned above decrease the soundness of the casting, i.e. decrease the density of the casting. After conducting several brainstorming sessions, the team members concluded that the density of the casting is the most important critical quality characteristic in the die-casting process as it is related to many internal defects (air entrapment, gas holes, porosity, shrink holes, etc.).

The objective of the die-casting process was to achieve 'better casting density' while minimising the effect of uncontrollable parameters. To have a clear picture of the process parameters affecting the density of casting, a 'cause and effect' diagram was constructed and is shown in Figure 9.23.

The cause and effect diagram shows that the most important process parameters that affect the casting density are piston velocity at first

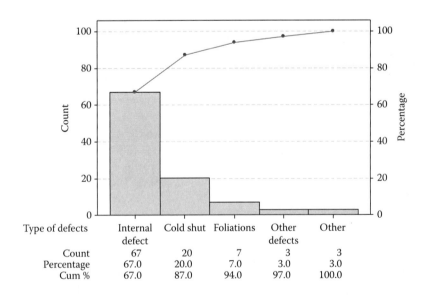

Type of defects	Internal defect	Cold shut	Foliations	Other defects	Other
Count	67	20	7	3	3
Percentage	67.0	20.0	7.0	3.0	3.0
Cum %	67.0	87.0	94.0	97.0	100.0

Figure 9.22 Pareto chart for the internal and external casting defects.

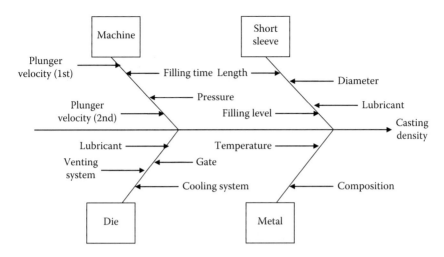

Figure 9.23 Cause and effect diagram for casting density problem.

Table 9.24 Process parameters with their ranges and values at three levels

Labels	Process parameters	Range	Level 1	Level 2	Level 3
A	Metal temp (°C)	610–730	610	670	730
B	Piston velocity 1st stage (m/s)	0.02–0.34	0.02	0.18	0.34
C	Piston velocity 2nd stage (m/s)	1.2–3.8	1.2	2.5	3.8
D	Filling time (ms)	40–130	40	85	130
E	Hydraulic pressure (bar)	120–280	120	200	280

stage, piston velocity at second stage, metal temperature, filling time and hydraulic pressure. From experience, it was revealed that non-linear behaviour of the parameters of the die-casting process can only be determined if more than two levels are used. The parameters along with their settings are given in Table 9.24. At this stage, it was essential to identify significant process parameters so that they are tuned properly to achieve the desired range of casting density.

9.5.3.4 Improve phase

In the improve phase, the team decided to carry out a designed experiment to identify the significant process parameters affecting the casting density. The most appropriate OA design to meet the experimental requirement is a 27-trial experiment (L_{27} OA), and the experimental layout is depicted in Table 9.25. The company was initially operating with the following settings:

$$A_1, B_1, C_1, D_2, E_3$$

The casting density is a 'larger the better' type of quality characteristic. Thus, the S/N ratio used is given by

$$\frac{S}{N} \text{ ratio} = -10 \ \log \left[\frac{1}{n} \sum_{i=1}^{n} \left(\frac{1}{y_i^{\,2}} \right) \right] \tag{9.1}$$

where $y_i =$ casting density for a trial condition.

Table 9.25 Results of L_{27} OA

Run no.	A	B	C	D	E	AB	AC	BC	y1	y2	y3	Average	S/N ratio
1	1	1	1	1	1	1	1	1	2.336	2.338	2.441	2.372	7.500
2	1	1	2	2	2	1	2	2	2.339	2.442	2.447	2.409	7.637
3	1	1	3	3	3	1	3	3	2.442	2.505	2.448	2.465	7.839
4	1	2	1	2	2	2	1	2	2.427	2.444	2.416	2.429	7.713
5	1	2	2	3	3	2	2	3	2.545	2.577	2.595	2.572	8.210
6	1	2	3	1	1	2	3	1	2.435	2.336	2.374	2.382	7.538
7	1	3	1	3	3	3	1	3	2.716	2.728	2.701	2.715	8.680
8	1	3	2	1	1	3	2	1	2.346	2.429	2.392	2.389	7.566
9	1	3	3	2	2	3	3	2	2.439	2.442	2.445	2.442	7.759
10	2	1	1	2	3	2	2	1	2.445	2.501	2.487	2.478	7.884
11	2	1	2	3	1	2	3	2	2.439	2.441	2.398	2.426	7.701
12	2	1	3	1	2	2	1	3	2.418	2.381	2.443	2.414	7.658
13	2	2	1	3	1	3	2	2	2.542	2.513	2.504	2.520	8.031
14	2	2	2	1	2	3	3	3	2.459	2.463	2.445	2.456	7.808
15	2	2	3	2	3	3	1	1	2.543	2.585	2.591	2.573	8.212
16	2	3	1	1	2	1	2	3	2.441	2.493	2.502	2.479	7.887
17	2	3	2	2	3	1	3	1	2.594	2.588	2.591	2.591	8.274
18	2	3	3	3	1	1	1	2	2.539	2.542	2.545	2.542	8.108
19	3	1	1	3	2	3	3	1	2.474	2.495	2.489	2.486	7.914
20	3	1	2	1	3	3	1	2	2.603	2.595	2.588	2.595	8.288
21	3	1	3	2	1	3	2	3	2.438	2.473	2.452	2.454	7.803
22	3	2	1	1	3	1	3	2	2.704	2.685	2.692	2.694	8.611
23	3	2	2	2	1	1	1	3	2.640	2.682	2.654	2.659	8.497
24	3	2	3	3	2	1	2	1	2.703	2.698	2.691	2.697	8.623
25	3	3	1	2	1	2	3	3	2.671	2.679	2.685	2.678	8.562
26	3	3	2	3	2	2	1	1	2.726	2.717	2.720	2.721	8.699
27	3	3	3	1	3	2	2	2	2.745	2.747	2.752	2.748	8.785

Table 9.26 Average values of S/N ratios for each process parameter at different levels

Process parameter	A	B	C	D	E
Level 1	7.827	7.803	8.087	7.960	7.923
Level 2	7.951	8.138	8.076	8.038	7.966
Level 3	8.420	8.258	8.036	8.201	8.309

Each trial condition was repeated three times (i.e. $n=3$). The S/N ratios are computed for each of the 27 trial conditions. The average values of the S/N ratios for each parameter at different levels for all the trials are listed in Table 9.26. The influence of interactions on the casting density was negligible based on the analysis and was thus omitted from further study.

From Table 9.26, it is clear that casting density is at maximum when the process parameters A, B, D and E are kept at level 3 and parameter C at level 1. Once the optimum settings of process parameters were identified, the team members decided to implement 5S system and total productive maintenance (TPM) to establish a clean environment within the shop floor and to reduce the idle time of machines and employees on the shop floor.

9.5.3.4.1   Confirmatory test   In order to validate the results obtained from the improve phase, a confirmatory experiment was performed using the optimal setting of process parameters A, B, D and E at level 3 and C at level 1. The average value of casting density was computed as 2.75 g/cc. This resulted in an increase of casting density by over 12%. In order to check that the results were valid and sound, it was decided to observe the value of casting density for the next 3 days of production.

9.5.3.4.2   5S system and TPM   Top level management decided to implement the 5S system in order to establish a standard approach to housekeeping within the organisation and help reduce the non-value-added time for employees (Womack and Jones 1996). Moreover, there was also concern about the health and safety issues of the employees as the average number of accidents on the shop floor were increasing per year.

The 5S training pillars were implemented on the shop floor, which helped the organisation in the following ways:

- A day-to-day floor cleaning programme was initiated, and it was ensured that the employees had sufficient lighting on the shop floor to work in the afternoon and night shifts.
- In order to minimise the idle time at each process, operators were provided with a rack to place the items correctly in the respective block provided in the rack.

- The trimming unit was moved nearer to the die-casting machine so that time was saved in transportation from die-casting machine to trimming press.
- The cleaning of dust particles, grease and oil from the machines helped to ensure the health and safety of employees.

The implementation of the 5S system helped to organise the work environment, standardise the workflow and assign clear ownership of processes to employees. It also helped in increasing the productivity by reducing idle time of some processes.

A TPM programme was introduced to the organisation in the late 1990s and was a complete fiasco due to lip service provided by management without them showing any interest in actual implementation of programme. TPM was only used for documentation purposes and for attracting customers. Tough competition within the market place forced management to rethink on proper implementation of the TPM programme within the Lean Sigma framework to markedly increase production and, at the same time, increase employee morale and job satisfaction. There was a constant increase in in-process inventory, machine downtime and idle time at work stations, which was easily tackled by proper implementation of TPM programme. The steps taken by management to facilitate effective implementation of TPM are listed below.

- Periodic maintenance of machines, i.e. cleaning, lubrication, inspection and corrective action on all machines on the shop floor
- Collection and analysis of data on downtime of machine and remedial action to increase the overall equipment effectiveness (OEE) and thus the overall plant efficiency (OPE)
- Creating an equipment improvement team and TPM area coordinators to monitor the proper implementation of the programme
- Involving employees at all levels of organisation to achieve zero defects, zero breakdown and zero accidents in all functional areas of the organisation
- Accentuating the training programme for effective implementation of programme

9.5.3.5 Control phase

The main purpose of the Six Sigma methodology is not only improving the process performance but also having the improved results sustained in the long run. Hence, the standardisation of the optimal process parameters setting is required. The die-casting process has been improved by optimising the critical process parameters A, B, C, D and E to around 730°C, 0.34 m/sec, 1.2 m/sec, 130 ms, and 280 bar, respectively. For measuring accurate values of the above process parameters, different sensors (pressure sensors, temperature sensors and position and velocity sensors) are used. The implementation of the aforementioned suggestions resulted in enhanced profitability of the organisation. X-bar and R control charts were used to make sure that the process is stable, and it is observed that none of the points have gone outside the control limits.

The management team has decided to implement a mistake-proofing exercise to prevent the occurrence of other types of defects in production. The following points have been taken into consideration while executing the mistake-proofing exercise:

- Checking the defects at the preliminary design phase so that defects are not passed to the production stage.
- FMEA, in-house scrap and rework data, inspection data and analysis of customer complaints were used to pinpoint potential problems that could be resolved by mistake proofing.
- Cross-functional teams were formed to discuss the manufacturing and design problems that are likely to cause mistakes/defects/ failures.
- Sharing of information related to company performance with its employees.
- Training people on the shop floor regarding details of production and quality issues as well as other activities such as problem solving and team building.
- Use of control charts and graphs at each processing stage to keep the employees aware of the real-time performance at the respective stages of production.
- To motivate and recognise employees' contribution in establishing best practices within the organisation.
- To reward and recognise the employees involved in the project.

9.5.4 *Typical benefits of the project*

The implementation of Lean Sigma methodology has helped the case study organisation in

- Reducing the machine downtime
- Establishing a standard housekeeping procedure
- Increasing the confidence level among employees
- Instigating a sense of ownership among employees
- Enhancing OEE
- Rectifying customer complaints
- Reducing inventory
- Reducing machine set-up time
- Reducing the number of accidents at the workplace

The savings generated by the organisation by achieving improvements in the aforementioned areas are as follows.

- The decrease in machine downtime from 1% to 6% helped in increasing the OEE. This resulted in estimated savings of over US $40,000/year.
- WIP inventory was reduced by over 25% and resulted in estimated savings of over $33,000/year.
- Standard housekeeping procedures helped to reduce the number of accidents at the workplace significantly. This reduced the amount of compensation the management needed to pay to injured employees (around $20,000 on average/year).
- The savings generated due to reduction in defects were estimated around $46,500/year.

Thus, there was an improvement of around $140,000/year in monetary terms for the company after implementation of the Lean Sigma strategy. Table 9.27 presents the significant improvements in the key performance metrics after implementation of Lean Sigma methodology. The key metrics used for comparing the results after implementing the Lean Sigma methodology include defect/unit (DPU), process capability index (C_p), mean and standard deviation of casting density, first time yield (FTY) and OEE.

It can be inferred from the table that there was significant improvement in the key performance metrics achieved by the company. This motivated the management for horizontal deployment of the Lean Sigma approach in other areas of the organisation such as transactional processes, service-related processes, etc., and to share the benefits generated with its employees.

9.5.5   Challenges, key lessons learned and managerial implications

For any continuous improvement programme, it is important to discuss the challenges encountered and key lessons learned from the execution

Table 9.27 Comparison of key performance metrics (before vs. after)

Key performance metrics used	Before improvement	After improvement
Defect rate	0.18 DPU	0.0068
FTY	82%	99.32%
Process capability index (C_p)	0.12	1.41
Process mean	2.45	2.75
Process standard deviation	0.069	0.0059
OEE	48%	83%

of the project. It provides valuable lessons learned from previous projects that should be taken care of while starting the new project. In this case, convincing top management was the most arduous task as management was not ready to compromise on production to improve the quality of the final product manufactured. The top management people felt that investing in quality means increasing the cost of production, which they cannot afford to do when faced with stiff challenges from competitors.

It is quite natural to encounter resistance from employees if you try to introduce and implement some new problem-solving methodology such as LSS. The employees of the organisation under observation thought that implementation of the new process improvement methodology could endanger their job opportunities and poor performance could result in them losing their jobs. This particular issue was discussed among the senior management team and later on, got corrected by top management, convincing the employees that their jobs would not be in danger and that they would be rewarded for better performance at the team and individual levels, if needed. This gradually boosted confidence in the employees and eventually they were ready to embrace the proposed Lean Six Sigma methodology in their processes.

Moreover, resistance from management was also noticed when the team had decided to implement the 5S system in the organisation in order to ensure proper housekeeping and to reduce accidents in the factory by ensuring a safer environment. The management thought that ergonomics would have no impact on the performance of the employee and, ultimately, production. The management teams were convinced by showing them the savings that can be generated if accidents are avoided 'right first time' (RFT) and how proper housekeeping can reduce the idle time of the operator and machine.

The company was using different problem-solving methodologies for different problems based on their experience, and quite often the root causes were never identified or derived by the team. No standard methodology was followed in the business for problem-solving scenarios, and this resulted in total chaos across the company on many occasions. Lean Six Sigma provided senior managers with a standard road map for tackling problems efficiently and effectively, and one of the senior managers commented that 'the best feature of this powerful methodology is the integration of problem solving tools within the five-stage methodology and the use of data to challenge many managers who constantly use their intuition and gut feeling for problem solving exercises'. The application of LSS has provided greater stimulus among many engineers and managers in the case study company, and this has resulted in more applications of this powerful problem-solving methodology in other aspects of the company such as finance, administration, supply chain, human resources and new product development processes.

9.5.6   Recap of tools used

The tools used during this case study:

Brainstorming
VOC analysis
Data collection strategy
Gage R & R study
Cause and effect diagram
Pareto analysis
Failure mode and effect analysis
Taguchi orthogonal array experiment
5S practice
Overall equipment effectiveness

9.5.7   Summary

The implementation of LSS provided an impetus for establishing best practice within the company. It has provided the case study organisation with a performance benchmark on which they could base future performance improvement initiatives. The optimal setting for the die-casting process has improved the casting density by over 12%. The financial savings generated from the project were approximately US $140,000/year, and this has created a momentum in the further applications of the methodology across the business. Among the challenges in many SMEs are the financial and manpower constraints. This demands the development of a standard LSS road map showing how to get started and the subsequent implementation and deployment guidelines. Chapter 4 of the book provides such a road map, which can be utilised by a number of SMEs with limited budget and manpower constraints.

References

Gijo, E. V., Bhat, S. and Jnanesh, N. A. (2014). Application of Six Sigma methodology in a small scale foundry industry. *International Journal of Lean Six Sigma* 5(2): 193–211.

Gijo, E. V. and Sarkar, A. (2013). Application of Six Sigma to improve the quality of the road for wind turbine installation. *The TQM Journal* 25(3): 244–258.

Gijo, E. V. and Scaria, J. (2010). Reducing rejection and rework by application of Six Sigma methodology in manufacturing process. *International Journal of Six Sigma and Competitive Advantage* 6(1–2): 77–90.

Kumar, M., Antony, J., Singh, R. K., Tiwari, M. K. and Perry, D., (2006). Implementing the Lean Sigma framework in an Indian SME: A case study. *Production Planning and Control: The Management of Operations* 17(4): 407–423.

Index